Laboratory Robotics
A Guide to Planning, Programming, and Applications

© 1987 VCH Publishers, Inc., 220 East 23rd Street, New York, NY 10010-4606

Distribution: VCH Verlagsgesellschaft mbH, P.O. Box 1260/1280, D-6940 Weinheim,
Federal Republic of Germany

USA and Canada: VCH Publishers, Inc., 303 N.W. 12th Avenue, Deerfield Beach, FL 33442-1705, USA

Laboratory Robotics
A Guide to Planning, Programming, and Applications

**W. Jeffrey Hurst
and
James W. Mortimer**

W. Jeffrey Hurst
Group Leader, Analytical Research
Hershey Foods Corporation
Technical Center
Hershey, Pennsylvania 17033-0805

James W. Mortimer
Regional Manager
Zymark Corporation
Hopkinton, Massachusetts 01748

Library of Congress Cataloging-in-Publication Data

Hurst, W. Jeffrey (William Jeffrey), 1948–
    Laboratory robotics: A guide to planning, programming, and applications.

    Bibliography: p.
    Includes index.
    1. Robotics. 2. Laboratories—Automation.
I. Mortimer, James W. (James Winslow), 1955–
II. Title.
TJ211.H87   1987   542'.1   86-32545
ISBN 0-89573-322-6

© 1987 VCH Publishers, Inc.

This work is subject to copyright.

All rights are reserved, whether the whole or part of the material is concerned, specifically those of translation, reprinting, re-use of illustrations, broadcasting, reproduction by photocopying machine or similar means, and storage in data banks.

Registered names, trademarks, etc. used in this book, even when not specifically marked as such, are not to be considered unprotected by law.

Printed in the United States of America.

ISBN 0-89573-322-6   VCH Publishers
ISBN 3-527-26675-5   VCH Verlagsgesellschaft

Foreword

These are truly exciting times to be a chemist. Applied chemistry is the key to innovative products in health care and pharmaceuticals, in industrial chemicals and plastics, in food and agriculture, and in energy. Well-trained scientists, engineers, and technicians, supported by modern laboratories, are required to develop such products and to efficiently manufacture them with high quality standards.

Our needs for chemical technology are growing faster than our supply of qualified personnel and the demand for laboratory and analytical support continues to increase. In addition, highly-trained people are often forced to perform repetitive laboratory procedures rather than delegate them to less skilled people—who may introduce errors.

Fortunately laboratory productivity is improving. Faster instrumental techniques, automatic sample injectors, and laboratory computers are increasing sample throughput without proportional staffing growth. The next step in this evolutionary process is automation of sample handling and sample preparation—hence a new technology called laboratory robotics.

We should visualize laboratory robotics as a systems concept rather than simply a remote controlled, mechanical arm. Laboratory robotics is an extension of programmable computers which permits computers to do physical work as well as process data. Typical laboratory robotics systems include several modular instruments that perform common laboratory operations in conjunction with a robot arm for sample handling and a microprocessor-based controller that translates laboratory procedures into specific instructions for each element of the system.

Although automation relieves people of boring and time-consuming tasks, it does require a one-time investment in detailed planning. Just as people must be trained to perform new tasks automated systems must be programmed for each application. Each time people change assignments their replacements must be retrained but automated systems need only be trained once. While people adapt to inadequate instructions they often produce inconsistent results; successful automation procedures consistent results but requires disciplined planning.

Ideally, automation can be implemented through direct adaptation of manual procedures. Rarely, however, is this totally possible. Most existing laboratory apparatus is optimized for human use and may be rather awkward in an automated environment. For example, we use paper clips in the office to manually fasten papers but we convert to staplers for automatic fastening. The key to successful automation then is to require and preserve proven chemistries during the conversion from manual to automated procedures but then be creative and flexible in the selection of apparatus and techniques.

These ideas may explain why the acceptance of industrial robotics is slower than originally expected:

> ...robots, most people thought, would have manipulators that swung like human arms, grippers that grasped objects like human fingers, and sensing abilities comparable to human senses. Made in the image of humans, these robots would soon outperform humans in cost and efficiency.
>
> This strategy hasn't worked. It was fatally flawed because it assumed that humans are optimally designed to perform manufacturing tasks and therefore deserve to be emulated. This is not true. Humans are designed to throw stones, pick berries and climb trees.*

In manual procedures samples are typically processed in batches because people perform best doing one step at a time. Computers, on the other hand, are able to keep time and simultaneously control multiple tasks, thereby permitting serialization of automated procedures. Almost all automated devices, including laboratory instruments, are serial. Au-

---

*Seering, W.P. Who said robots should work like people? *Technology Review*, April 1985, p 59–67.

tomatic injectors with carousels simply convert from manual batch processing to automated serial operation. Serialized automation usually leads to higher throughput, greater human productivity, and more uniform results in comparison to manual batch processing.

All automation ultimately requires human participation. The goal should not be to remove people totally from the process but to integrate people and instruments into an efficient system. Certain tasks requiring judgement or complex eye-hand coordination are best done by people. Ideally, these human operations should be grouped together either before or after the automated steps.

New technologies need pioneers to adapt and apply them to real procedures. These pioneers become the teachers and support resources for subsequent users. The widespread use of new technologies provides a growing base of trained scientists and engineers. While reading this book and building your personal education in laboratory robotics, always remember: proven chemistry, disciplined planning, and creative implementation.

F.H. Zenie
Hopkinton, Massachusetts

Preface

In the past three years the concept of laboratory robotics has evolved from a curiosity to a phenomenon. Three years ago there was one robot manufacturer; this past year saw a dramatic increase in the number of robot manufacturers, as well as the emergence of vendors providing specific robot-supportive products and the development of cooperative agreements between robot vendors and corporations. We hope that this book will provide researchers and other interested parties with the ability to initiate an exploration of lab robotics.

This volume is not intended to be a lengthy treatise on robotics or lab automation, nor is it intended to discuss the many specific applications of robotics, as there are already several volumes on these topics. Our intent is to provide an introduction to the concepts, the language, and the reasoning behind the purchase of a lab robot along with some general applications of robotics, and a bit of crystal-ball gazing.

Acknowledgments

The concept for this book was generated by the Zymark Corporation (Hopkinton, Massachusetts), where we were introduced to lab robotics. We are deeply indebted to Zymark for their permission to borrow freely from their published and unpublished information. Certain individuals at Zymark deserve our special thanks. These include Burleigh Hutchins, creator and father of the Zymark Robot, and Frank Zenie, whose ability to express the concepts of robotics has given us extensive help and guidance in this venture. Additional thanks are also due to the applications and marketing staffs of Zymark, as the unique applications of robotics would not have been possible without these individuals. We also wish to thank the following organizations and individuals for their support and assistance: Perkin-Elmer Instrument Corporation; Microbot; IBM; St. Charles Manufacturing Company; Fisher Scientific Corporation; Analytichem International; and Hershey Foods Corporation, especially Kevin Synder, for drafting assistance, and Mildred Sholly, for typing the manuscript. Thanks are also due to our families, who put up with late nights, reminded us of impending deadlines, and lived through the dreaded writer's block. We owe a debt of gratitude to all these people and many others. We gladly give credit to all those who helped make this volume possible, while maintaining that any shortcomings are solely the responsibility of the authors.

The book concentrates on Zymark and many of their products with no apologies, because they introduced us to robotics and lived through the good and bad times associated with a new product and a new technology. While we have concentrated on the Zymark Robot, the concepts outlined in this text can be applied to any type of laboratory robot.

W. Jeffrey Hurst
James W. Mortimer

Contents

Chapter 1
The Need for Robotics in the Laboratory   1

Chapter 2
Automation through Laboratory Unit Operations   15

Chapter 3
The Justification of a Laboratory Robot   25

Chapter 4
Applications for Laboratory Robotics   39

Chapter 5
Getting Started in Laboratory Robotics   73

Chapter 6
Programming the Robot   93

Chapter 6
The Future of Laboratory Robotics   109

Glossary   119

References   125

Index   127

1
The Need for Robotics in the Laboratory

One of the newest tools in the analytical laboratory is the laboratory **robot**. This first chapter will explore the various reasons for the acquisition of a lab robot. As one might envision, the reasons are wideranging. This volume will be restricted to the topic of robots in the laboratory, but some of the reasons for the acquisition of a robot in the laboratory might also be applicable to other areas. This volume will not deal with robots and their widespread use in manufacturing operations as this topic has been well covered by others, and is well beyond the scope of this modest text.

The concept of robotics has been with us for a considerable amount of time. It has evolved from the concept of the automaton. The word automaton is derived from the Greek and means "having motion within itself." The term robot is derived from the Czechoslovakian word *robota*, which means slave, servant, or compulsory labor. It was used in 1921 in a play authored by Karl Capek and presented in Prague. The play was entitled *R.U.R.*, which stood for Rossum's Universal Robots. In this play, robots were invented to save humans from work but they developed a distaste for the imperfections and frailty of humanity and eventually took over the world. In 1926, director Fritz Lange introduced the public to the robot Maria in the movie *Metropolis*. Maria was another robot created by man and she eventually incited the workers to revolt. In 1939, Isaac Asimov started to write stories about robots with a built-in set of safeguards. These safeguards were eventually formalized into **Asimov's Three Laws of Robotics:**[1]

1. A robot may not injure a human being.
2. A robot must obey orders given to it by humans except where such orders conflict with the first law.
3. A robot must protect its own existence as long as such protection does not conflict with the first two laws.

As time progressed the public was introduced to other robots, including Robbie the Robot in *Forbidden Planet*, the robot of the television series "Lost in Space," and the most recent additions to the genre, R2-D2 and C-3PO of *Star Wars*. In the 1960s we were introduced to HAL of the spaceship Discovery in *2001: A Space Odyssey* and in the 1980s to SAL in *2010: The Year We Make Contact*. There is even a hobby robot produced by the Heath Company that is called HERO.

Robotics is therefore not new, and many who are first introduced to lab robotics expect a lab robot to resemble one of these fictional creations.

The Robot Institute of America defines a robot as "a reprogrammable multifunctional manipulator capable of moving material, parts or tools through variable programmed motions for the performance of a variety of tasks."[2] Others have defined it as a self-governing, programmable electrochemical device used by science and industry to perform a task faster, more cheaply, and more accurately than a human being. Some feel that robotics is essentially a sunset of automated instruments that populate many current labs. An automated operation allows all operations to be the same, whereas in robotics there is the possibility to **reprogram** the unit. Additionally, a robot is not limited to a single task. It can be classified as soft automation or flexible automation. A modern laboratory robot is a highly specialized, computer-driven programmable unit capable of accomplishing a variety of laboratory operations. Recently *Forbes* has dubbed the robot the "one armed chemist."[3] Figures 1-1 through 1-4 show pictures of laboratory robots.

In the area of laboratory robots, the Microbot Alpha saw sporadic lab usage prior to the first successful commercial robot, the Zymate, introduced by the Zymark Corporation in 1981. The Perkin-Elmer Masterlab System was introduced in early 1985 and the Fisher Scientific entry Maxx 5 was introduced in early 1986. Several other vendors are introducing robotic units for discrete operations such as automated weighing. Additionally, the Zymate unit has been used in several cooperative arrangements with vendors such as Hewlett Packard and one would expect that this trend would continue with other robot vendors. Figures 1-5 and 1-6 show line drawings of one of the current Masterlab robots available through Perkin-Elmer.

THE NEED FOR ROBOTICS IN THE LABORATORY

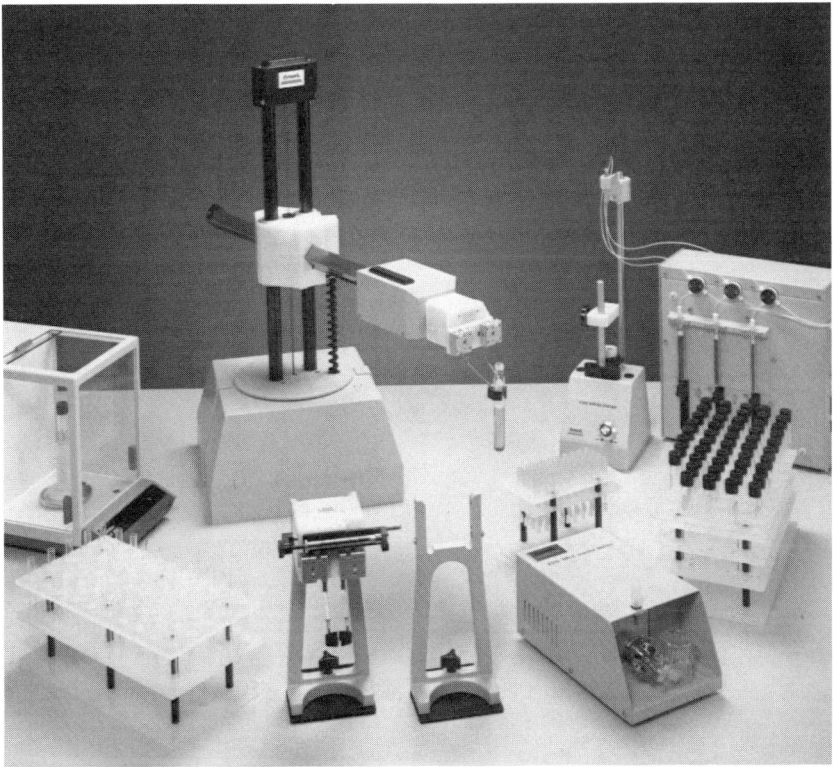

FIGURE 1-1. Zymark robot. (Photo courtesy of Zymark Corporation.)

Now that one has been introduced to the concepts of robotics, the question might arise as to why an analyst would want to acquire such a unit. This chapter will explore a variety of reasons involved in the need for robotics in a laboratory. Table 1-1 summarizes some data on these robots.

Several of the first reasons are not what one would consider sound, but have been overheard in discussions with many people involved in lab operations. Some might consider purchasing a robot so their lab would be in the forefront of the newest technology. This type of reasoning also results in problems for both the buyer and seller of the instrument and should be disregarded.

Another form of this type of reasoning is the need to keep up with other labs. This can be compared to "keeping up with the Joneses."

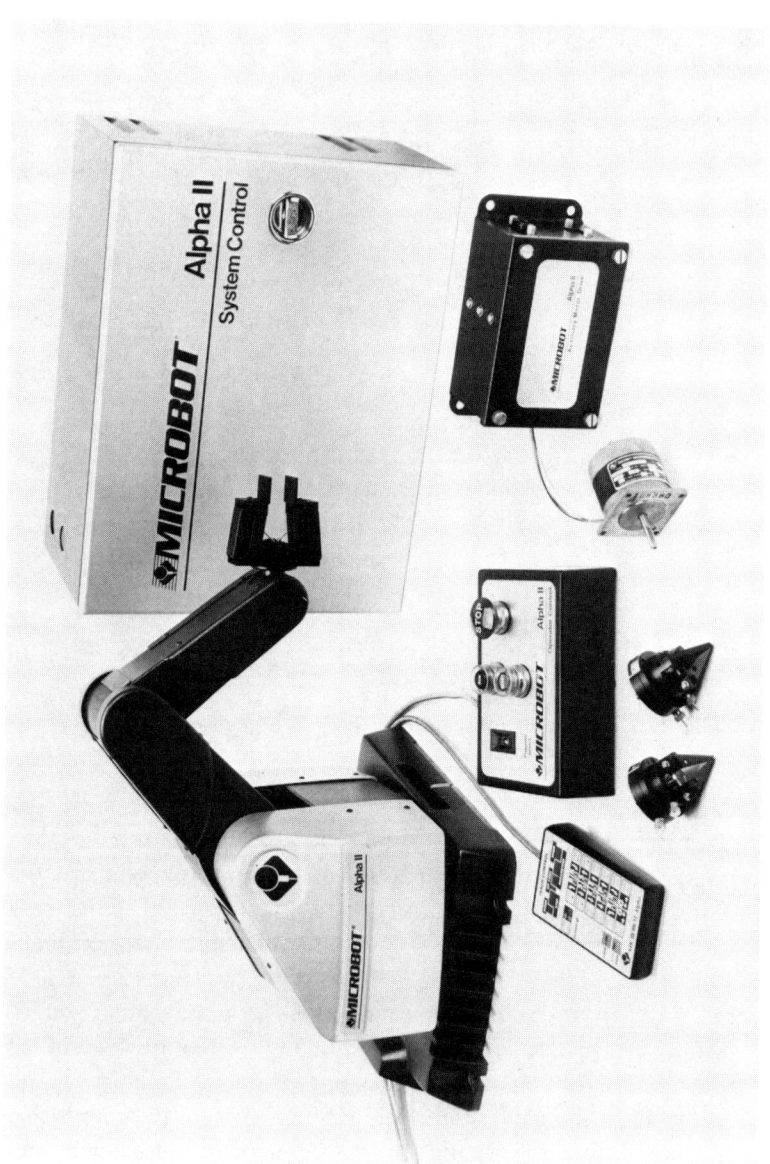

FIGURE 1-2. Microbot Alpha II robot system and accessories. (Photo courtesy of Microbot Inc.)

THE NEED FOR ROBOTICS IN THE LABORATORY

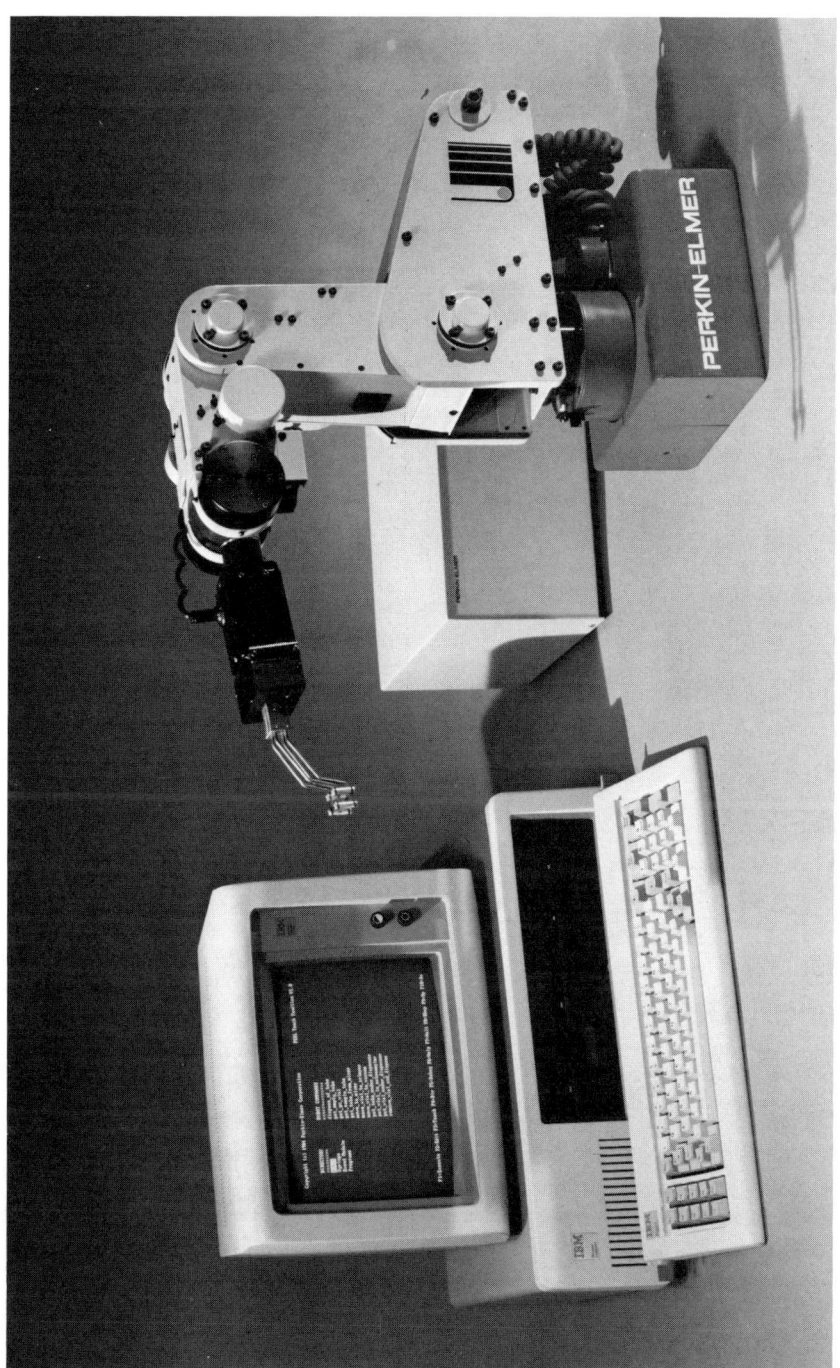

FIGURE 1-3. Perkin-Elmer laboratory robot. (Photo courtesy of Perkin-Elmer Corporation.)

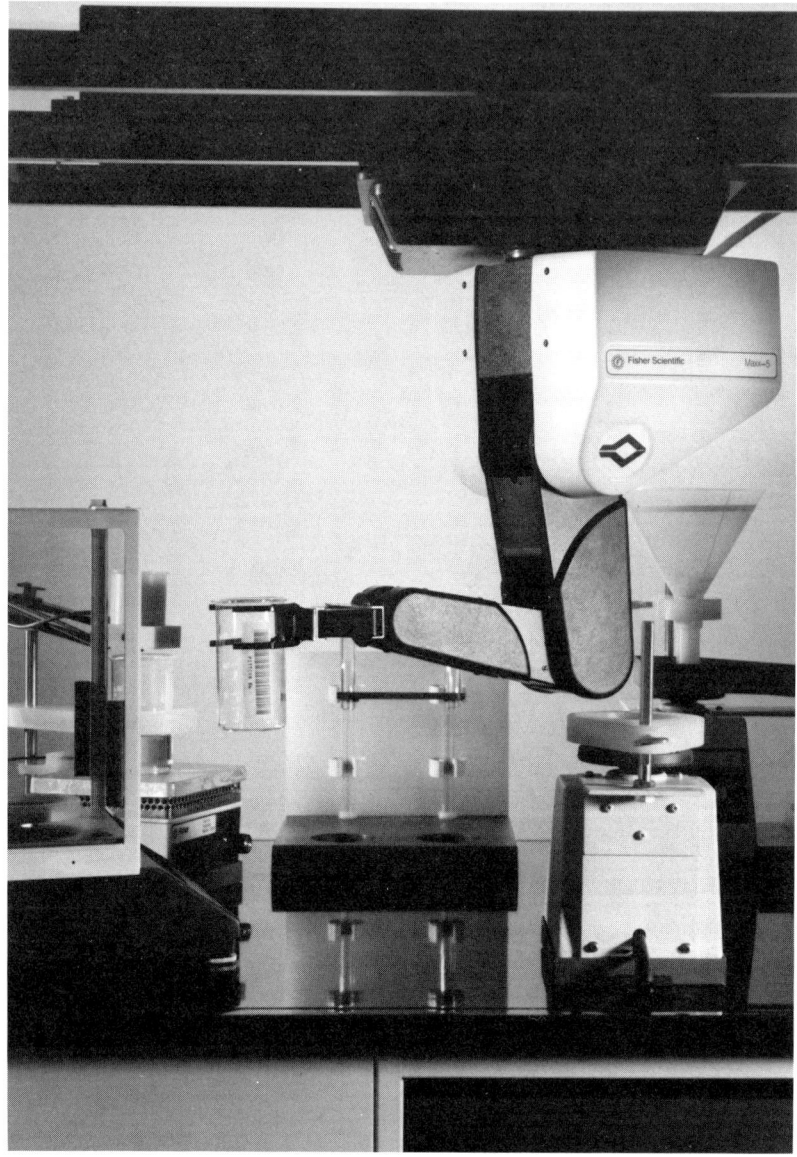

FIGURE 1-4. Maxx 5 robot. (Photo courtesy of Fisher Scientific.)

THE NEED FOR ROBOTICS IN THE LABORATORY

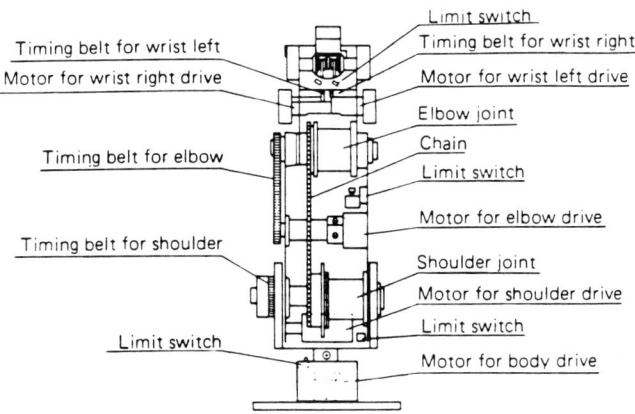

FIGURE 1-5. Interior of arm for Masterlab robot. (Courtesy of Perkin-Elmer Corporation.)

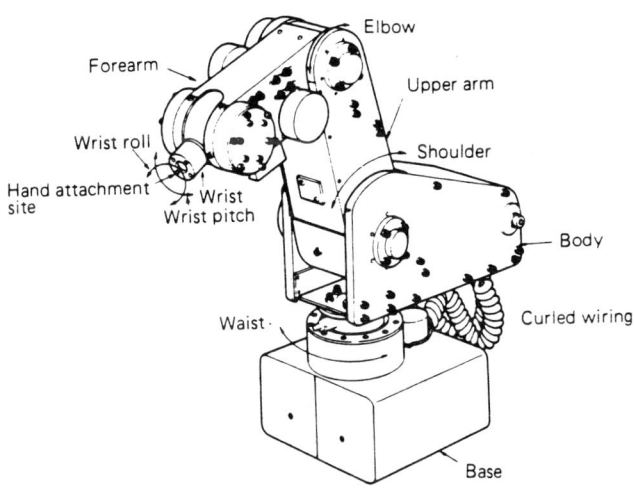

FIGURE 1-6. Outer appearance of Masterlab robot. (Courtesy of Perkin-Elmer Corporation.)

TABLE 1-1. Comparison of Some Available Laboratory Robots

Robot type	Coordinate system	Operation range	Programmable speeds	Reach	Lift capacity (kg)	Positioning accuracy (mm)	Drive system	Computer interface	Controller	Teaching	Miscellaneous
Zymate	Cylindrical	Base 360°	*	60 cm	2	2.5	Cable driven with PC Servo motors and potentiometers as position encoders	RS232 Serial	Integral	Soft keys assoc. with controller or teaching pendant	Has ability to change hands
Microbot Alpha II	Revolute	Base 345° Shoulder 145° Elbow 135° Pitch 180° Roll 540°	Yes	46.7 cm	1.362	0.5	Stepper motors	Dual	6502A Microprocessor	Keyed or via host computer	Standard fingers
Perkin-Elmer (Mitsubishi Movemaster) RM501	Revolute	Base 300° Shoulder 130° Elbow 90°C Wrist Pitch 90° Wrist Roll 180°	Yes	66 cm	1.2	0.5	Stepper motors	RS232C	IBM PC	Host computer	Standard fingers

*A faster version of the Zymate Robot has been introduced

At scientific meetings, discussions of lab robotics always crop up and reasons for purchasing a robot are discussed. Several scientists have been overheard to say, "We'll probably buy a lab robot because my boss doesn't want to appear to not be current." While this situation might seem ridiculous, it does occur and purchases for this reason should be avoided. This is probably one of the worst reasons for a purchase of this magnitude since it will most likely result in unrealized expectations and general disappointment.

If these two reasons can be considered the worst reasons to become involved in robotics, then the following item can be considered to be one of the best reasons for a lab's involvement in this technology. Robotics fills the missing link in sample preparation. Current laboratories tend to have an extensive inventory of automated analytical instruments ranging from microprocessor controlled pH meters to intelligent autosamplers and dedicated autoanalyzers. While these instruments do contribute to lab productivity, they do not allow for the automated preparation of samples. Many of these modern instruments have thruput allowing several hundred determinations to be completed in an hour. This can result in labs devoting substantial time to sample preparation so that autosamplers can be loaded. In many labs, people can spend all day preparing samples so they can be run overnight. Such batch processing can result in instrumentation sitting unused for substantial amounts of time while samples are being prepared. This type of operation is unproductive from equipment and personnel aspects. Robotics can fill this area of the missing link in sample preparation. Figure 1-7 shows the three elements of a chemical analysis. There has been considerable progress in the latter two areas, but progress in sample preparation has not kept up. This is the area that is in need of most improvement.

When human analysts accomplish an analytical assay or task, each accomplishes that task in his own unique way. Each lab has a laboratory manual or methods manual that outlines each assay normally done in that lab. This manual will list equipment, reagents, and procedures that are necessary to complete a particular assay. It is most likely that each ana-

| Sample preparation | Analytical measurement | Data reduction and documentation |

FIGURE 1-7. The three elements of a chemical analysis.

lyst running an assay has his own copy of the method rigorously annotated with peculiarities of that assay that are not part of the method write-up. These annotations might typically include order of addition of reagents, a particular brand of reagent that seems to work better than another brand, and maybe a particular instrument to use. These are not major items and reflect personal preference, but they do result in the "underground laboratory" that operates in many labs.

While these types of lab organizations seem to have no visible effect on lab operations they do tend to create problems. For example, when two analysts are required to accomplish the same assay, they tend to complete the given task in two different manners, sometimes producing widely differing results because each tends to keep his own set of notes for the assay. These notes are usually not entered as changes into the official method because they are thought of as minor modifications and unworthy of inclusion on official documents.

A laboratory robot tends to standardize these operations and remove them from the underground laboratory. With robotics, a method can be developed and the analysis can be carried out in the same manner on a 24 hour-a-day, 7 day-a-week basis. In fact, an assay that is done on only an occasional basis would also be suitable for robotics, since once the method is developed it is **archived** for further use at a later date. Archiving eliminates dependence on a single person in the lab who accomplishes that particular assay.

A laboratory robot is less likely to commit manipulative errors than humans. This goes along as a corollary to the earlier comments about the elimination of the weak link in sample preparation. A properly programmed and configured robot will always accomplish the same task in the same manner. It will not deviate from the predefined program that is was previously taught.

There is increasing public concern about exposure to toxic chemicals and environments. The general public is becoming more aware of exposure to potentially harmful environmental agents and is rightly concerned about potential hazards. Some of these hazards include exposure to low levels of ionizing radiation and harmful chemicals associated with manufacturing environments. There also is concern to keep this exposure ALARA (As Low As Reasonably Allowed). This concern must be balanced against the valid wants and desires of society. For example, while everyone is concerned about the toxic chemicals used in some manufacturing operations, they would not want to live without the various consumer goods produced by some of these operations. An extremely large segment of the population owns a car yet we are told that we should shift our transportation to the use of mass transit to cut down on air pollution. An informal observation would indicate that we do not want to give up our personal transportation. This example is given to illustrate the trade-offs that are evident in our modern society. We want to decrease our exposure to potentially toxic chemicals and environments but do not want to sacrifice our consumer wants. This is an area that provides a potential perfect fit for the modern lab robot, as it is capable of working in potentially harmful environments, thus eliminating human exposure, and it is also capable of operating in low-level radiation environments. One might expect that such a unit will be configured with the proper shielding to allow it to operate in higher level radiation fields and in corrosive environments.

The current generation of lab automation is dedicated to stand-alone analyzers that are capable of doing a few discrete assays. While this number is extremely large in some cases, as in the HPLC and FIA systems, these units are nevertheless limited in their flexibility. Additionally, only those applications that are reasonably fixed or unchanging can justify the expense in stand-alone equipment. A modern multifunctional laboratory robot will allow an extremely large amount of flexibility in a laboratory. In his presentation to the Analytical Laboratory Managers Meeting, Frank Zenie[4] defined flexible automation as programmed by the user to perform multiple, similar procedures and capable of being quickly reprogrammed to accommodate new or revised procedures. He outlined this philosophy by using the grid arrangement shown in Figure 1-8 wherein flexible automation fits into the area above manual techniques but below specialized automated equipment.

A laboratory robot will add to increased lab safety, by the potential elimination of exposure to harmful chemicals. Additionally, when properly programmed it would cut down on broken glassware and other maladies of this sort. Sometimes accidents are attributable to tired or improperly trained personnel. A lab robot is trained once and does not become tired or have a limited attention span.

A robot makes an important contribution to the lab in the area of quality control. This contribution is really a compilation of the various reasons for lab robotics that we have previously mentioned. There is less

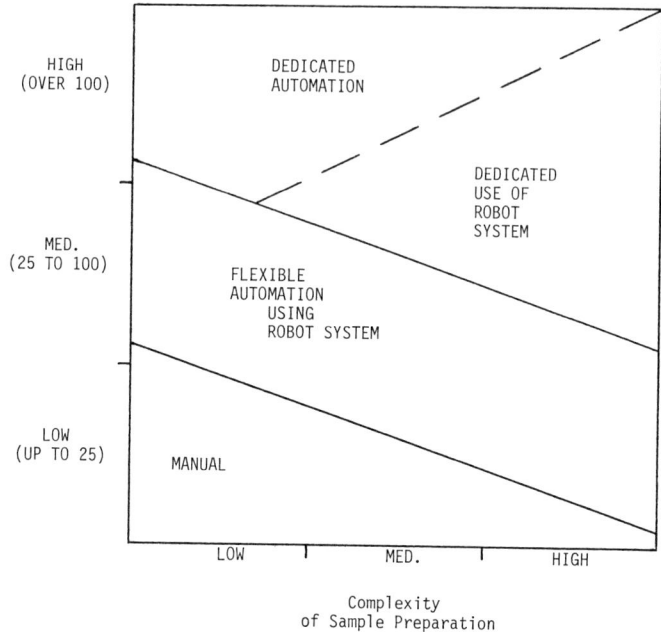

FIGURE 1-8. Comparison of samples/day vs. sample complexity.[4]

chance for manipulative errors, more safety, and, when the robot is equipped with a printer, less chance of transcription errors when transferred data is transferred from worksheets to a final document.

In our society the mention of robots first conjures up visions of R2-D2 and then visions of people out of work. The lab robot is far from both visions. The use of such a unit will tend to create a more invigorating working environment. Laboratories can automate their routine assays and allow their technical staff to become involved in more creative activities. This creates a twofold advantage: the automation of routine assays will result in increased productivity and it will also create a much more rewarding work environment by eliminating repetitive assays. This allows for better utilization of a technical staff that is likely to be in short supply.

Other advantages of the laboratory robot over a dedicated machine include the fact that it is easily programmable and therefore versatile. Laboratory robots allow an increase in efficiency because they are used continually, with samples loaded and unloaded throughout each analysis.

Additionally, each sample now has a uniform sample history since each is assayed in an identical manner. When samples are loaded into an autosampler or discrete analyzer, the analyst is usually required to wait until all assays are done in order to obtain data on the first sample. Robotics assures ready availability of results since samples are done in a serial rather than a batch mode.

This chapter has given only some general guidelines for the need for lab robotics and is not all-encompassing. Each lab will have its own unique reasons for the acquisition of a lab robot that will support the activities of that laboratory.

2
Automation through Laboratory Unit Operations

If all laboratory procedures are different, how can they be automated? A researcher should look for the similarities or the common elements in lab procedures and not focus on the differences. Just as groups of various amino acids join together in different combinations to form peptides, which in turn combine to form proteins, sequences of common steps become laboratory operations, which build upon each other to form a complete laboratory procedure.

Most laboratory operations consist of sequences of common steps. These building blocks are called laboratory unit operations (LUOs). It is through some combination of these LUOs that any laboratory procedure may be outlined. Ten classes of LUOs are possible[5] and these classes may have subclasses (for example, liquid handling includes pipeting, diluting, and dispensing).

LABORATORY UNIT OPERATION CLASS AND DEFINITION

The following is a list of the ten classes of LUOs with their definitions:

1. *Weighing*. Quantitative measurement of sample mass. The automation system must be capable of interfacing with common laboratory electronic balances,

i.e., opening the balance door, taring a suitable container, delivering the desired amount of sample, and recording the actual weight of the sample (Fig.2-1)
2. *Grinding.* Reducing sample particle size.
3. *Manipulation.* Physical handling of laboratory materials, i.e., the transfer of samples from a rack to the nozzle of a diluter station or to a sipper on a spectrophotometer. Pouring, capping, and crimping also fit into this category (Fig. 2-2).
4. *Liquid handling.* All physical handling of liquids—reagents and samples. The automation system must be able to handle and control the necessary stations to perform the desired pipetting, dispensing, and diluting called for in an application (Figs. 2-3 and 2-4).
5. *Conditioning.* Modifying and controlling the sample environment. Subclasses could include some of the following: shaking (linear, orbital, vortexing), heating, cooling and atmospheric blanketing.
6. *Separation.* Coarse mechanical and precision separations. Several subclasses that fit into this category include: solid phase extraction, liquid/liquid extraction, precipitation, filtration, and centrifugation (Fig. 2-5).
7. *Measurement.* Direct measurement of physical properties. The automation system must be able to interface with typical laboratory measuring equipment such as pH meters, spectrophotometers, conductivity meters, temperature probes, and liquid of gas chromatography detectors.
8. *Control.* Use of calculation and logical decisions in laboratory procedures. Based upon the input received from measuring devices or calculations performed, a decision is made and action taken. An automation system has limited value and flexibility if it cannot receive input and alter its behavior based upon that information.
9. *Data reduction.* Conversion of raw analytical data to usable information. Some examples are peak integration, spectrum analysis, and molecular weight distribution.
10. *Documentation.* Creating records and files for retrieval. In an automation system, the way in which each sample was prepared can be documented or the information can be transmitted to a host computer.

Any laboratory procedure can systematically be specified in terms of: (1) the required LUOs, (2) their sequence in the procedure, and (3) what happens to each LUO.

LUOs are the building blocks for all laboratory scale operations. They can be combined in many arrangements to meet specialized requirements. An automated laboratory system capable of performing various LUOs provides the flexibility for changing needs. One might examine several common analytical procedures, break them down into their respective LUOs and then discuss how best they could be automated.

A manual method for inductively coupled plasma spectroscopy (ICP) analysis of lubricating oils could involve the following:

AUTOMATION THROUGH LABORATORY UNIT OPERATIONS

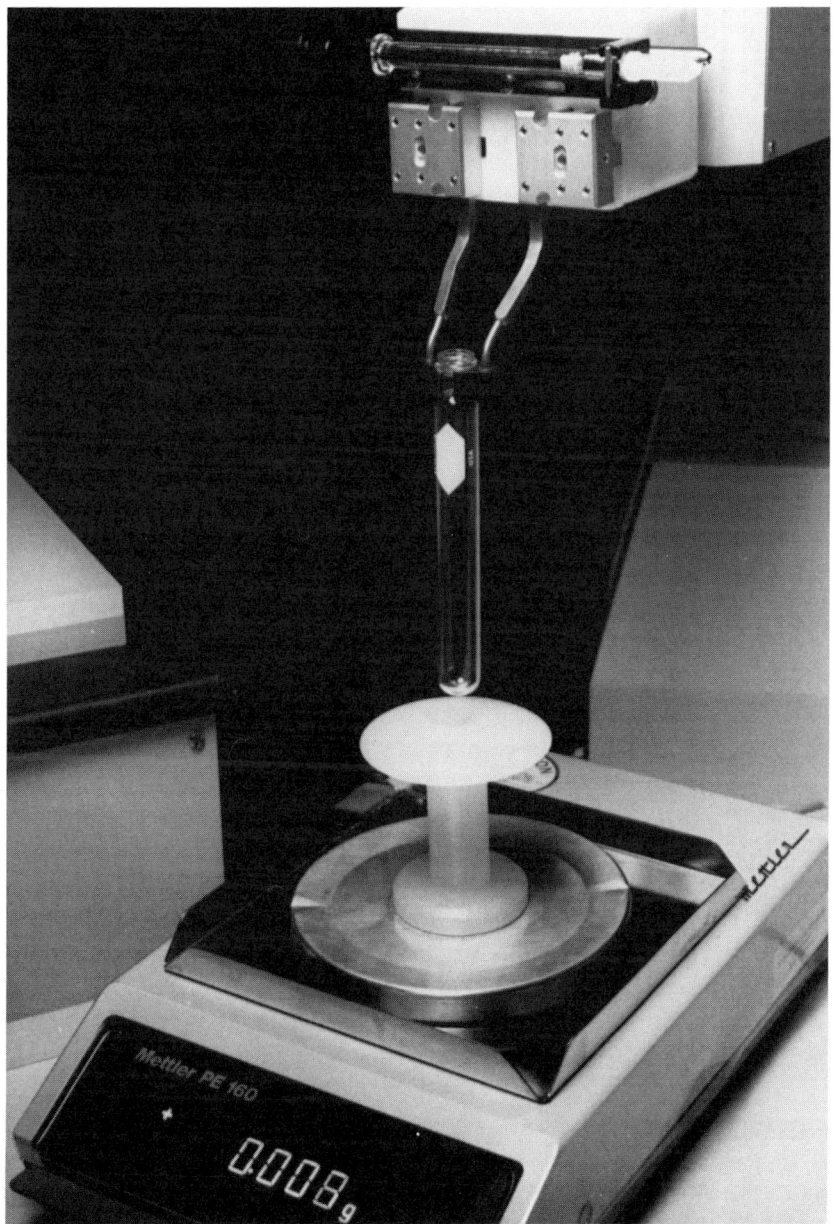

FIGURE 2-1. Robot weighing tube on balance. (Photo courtesy of Zymark Corporation.)

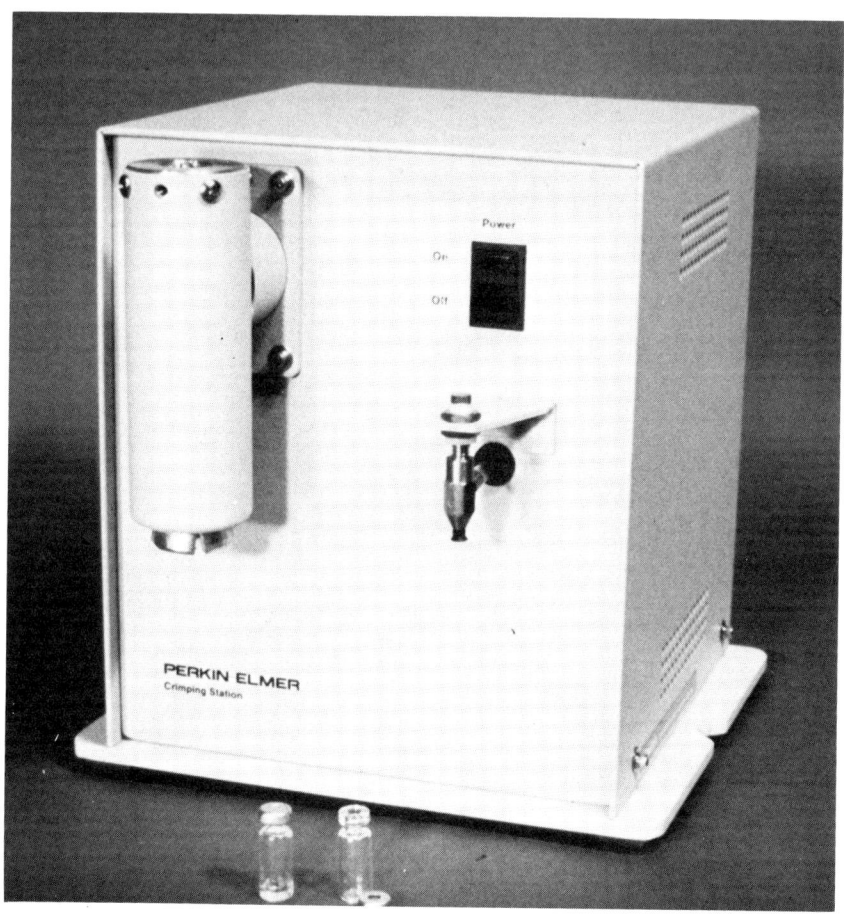

FIGURE 2-2. Crimping station used in laboratory robotics. (Photo courtesy of Perkin-Elmer Corporation.)

AUTOMATION THROUGH LABORATORY UNIT OPERATIONS 19

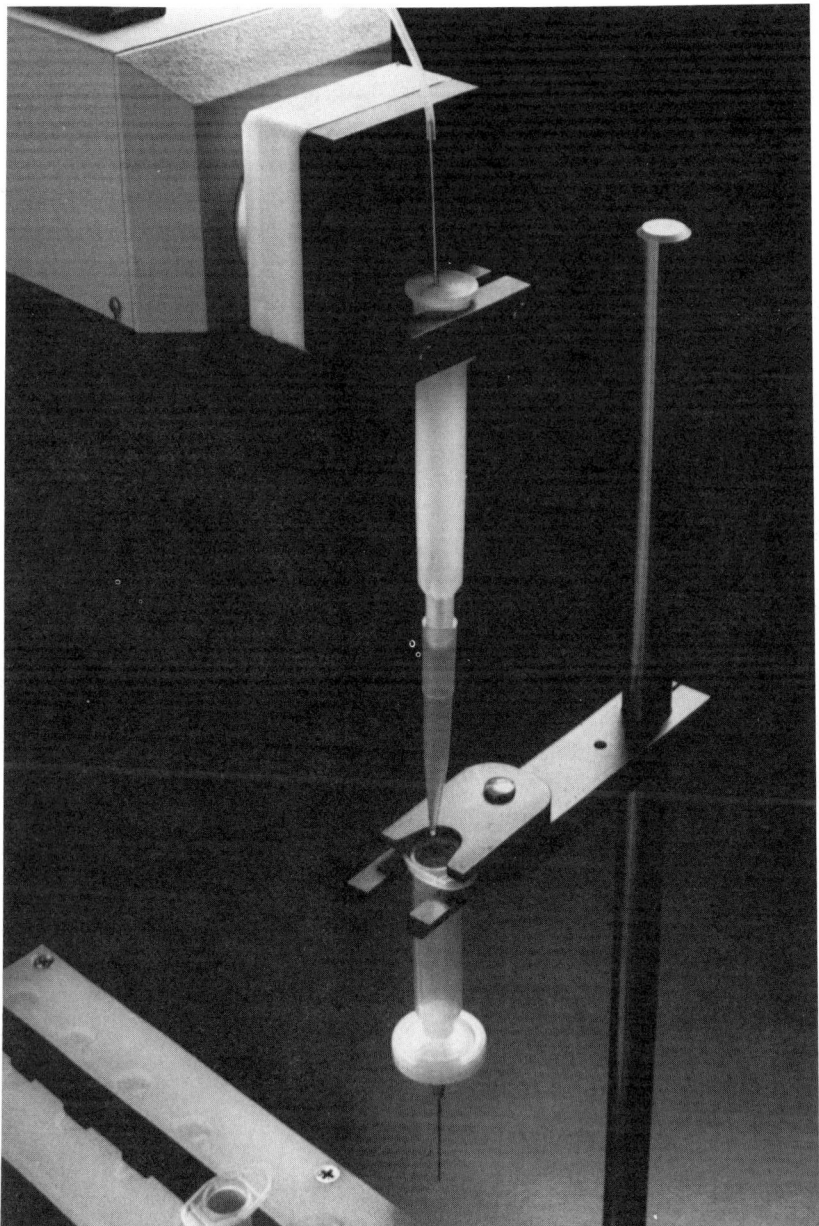

FIGURE 2-3. Robot dispensing sample into syringe for filtering. (Photo courtesy of Zymark Corporation.)

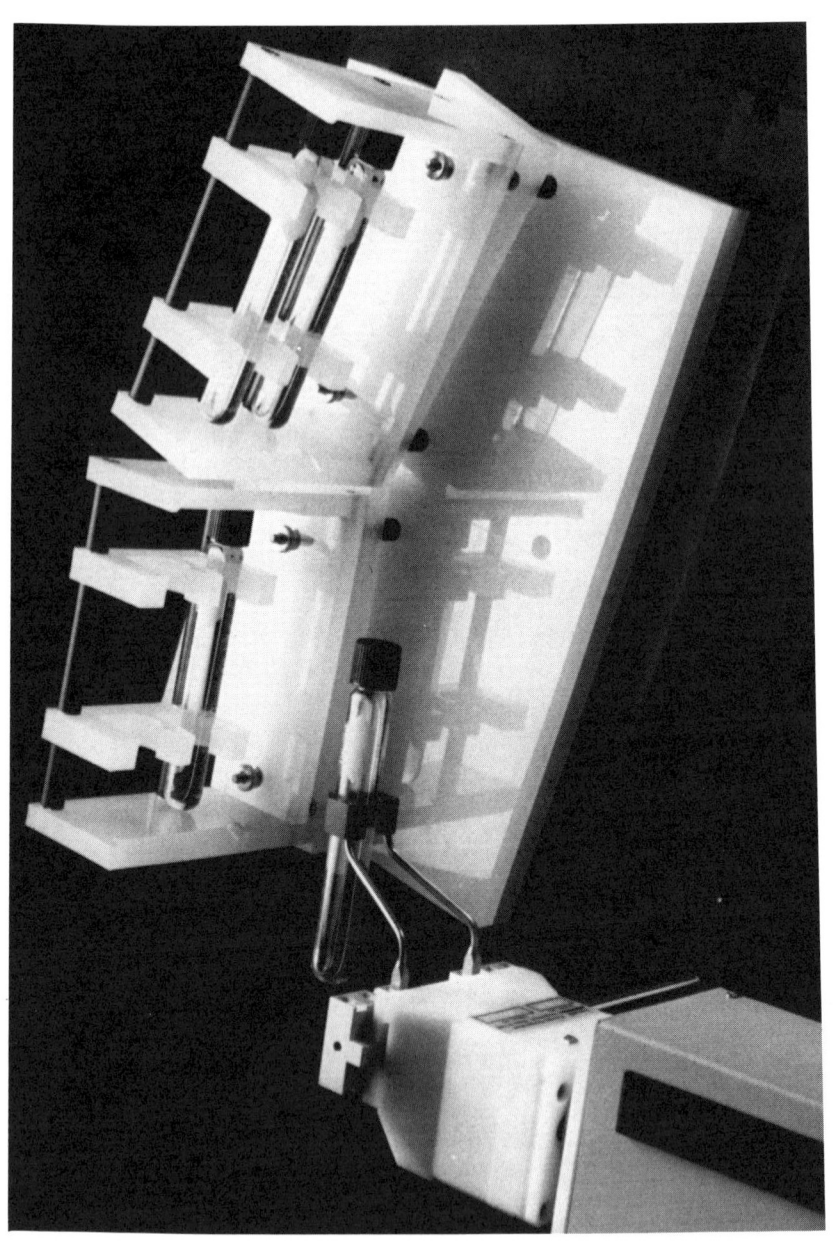

FIGURE 2-4. Robot placing test tube on conditioning station. (Photo courtesy of Zymark Corporation.)

AUTOMATION THROUGH LABORATORY UNIT OPERATIONS

These steps can be broken down further into LUO sequences:

1. Prepare the sample

Step	Objective	LUO	Procedure and parameters
1	Transfer sample to test tube	Liquid handling	Pipet 1 ml of sample; dispense into test tube
2	Introduce internal standard	Liquid handling	Dispense 1 ml of internal standard
3	Dilute sample	Liquid handling	Dispense 8.9 ml of solvent
4	Mix sample	Condition	Vortex for 5 sec

2. Condition the column

Step	Objective	LUO	Procedure and parameters
1	Transfer column to work station	Manipulation	Pick up column from rack; place in work station
2	Wet column	Liquid handling	Plug into column; dispense 5 ml of methanol; push through column with air

3. Add sample to column and elute to waste

Step	Objective	LUO	Procedure and parameters
1	Transfer sample to column	Liquid handling	Pipet 5 ml of sample; dispense into column
2	Push through column	Separation	Plug into column; push through column with air; collect waste

4. Elute the sample

Step	Objective	LUO	Procedure and parameters
1	Move column from work station to carousel	Manipulation	Remove column from work station; move to carousel
2	Introduce elution solvent	Separation	Dispense 5 ml of solvent A; push through column with air; collect sample in vial

Many times the manual procedure to be automated has not been fully documented. *Take the time to do this before attempting to automate.* The approach most desired in documenting any application is to view it as a combination of well defined LUOs.

3
The Justification of a Laboratory Robot

When a scientist decides to make a large or unique purchase of equipment for the lab, he is ultimately faced with the prospect of justification. This should not be surprising because with any purchase, large or small, for home or work, some sort of justification is necessary. This justification can range from a relatively informal mental justification, as to why you might need a new stereo, to a formal justification procedure involving forms, letters of justification, and signoffs by corporate officers for a large equipment purchase. When one has made the initial decision to purchase a laboratory robot the justification phase is most critical and, in fact, if this is the first purchase of this sort for an organization, this phase is most critical. Most likely if this first opportunity is ruined, it will be the only opportunity of this type offered to an individual. The discussions in this chapter will concentrate on these first purchases, since many subsequent purchases are additions to an ongoing robotic operation. It will discuss the factors that must be considered when going through the justification of a laboratory robot. These factors include not only capital but labor and implementation costs as well.

In his article on robots and lab automation, Liscouski[6] states that automation must be a means to achieve a goal, a process for solving some laboratory problem. If there is not a clearcut case for robotics in a laboratory, it should not be considered. What a scientist doesn't need is an expensive piece of equipment that ends up becoming the lab paperweight, since that will ruin additional opportunities for subsequent purchases that might have excellent opportunities for success. This is not to say that many first purchases were not made by central research and development groups to evaluate the fate of robotics in their organizations. This is indeed a good reason for a first purchase—the evaluation of a new technology. When high-performance liquid chromatography (HPLC) was the new technology, many organizations purchased an HPLC to evaluate this new technology on a limited number of samples, and now HPLC has a firm, premier place in analytical laboratories.

We are currently being flooded with a vast amount of information about productivity in business. Productivity is the word of the 80s. Businesses are not really asking their employees to do more work with less but to "work better." This is the key to productivity. One of the major justifications for the purchase of a laboratory robot is its ability to increase productivity. In his tutorial on robotics, Dessey[7] pointed out that the cost of human labor approaches $20/hr where a robot costs $5/hr, so potential increases in productivity are obvious.

Anyone who has seen a modern laboratory robot will soon notice that it does not work at an exceedingly fast pace but rather at a regular pace. It does have the ability to work longer with fewer mistakes. This concept of productivity has three facets: time, people, and money.

In the area of time, it has been mentioned earlier that a robot works longer. It does not take lunch breaks or coffee breaks, it doesn't get ill or have to answer the phone. When properly programmed it has the ability to work 24 hours a day, 7 days a week, and 365 a year. In his article on trends in lab automation, Frank Zenie,[4] states that this is the ability to work unattended and to work in extended operation (i.e., more than eight hours a day).

The purchase of a robot allows a manager the ability to make better use of his staff. All laboratories have a finite number of staff members. A robot allows many repetitive tasks to be fully automated, allowing a "cradle to grave" cycle of analysis for a sample. The automation of these repetitive, sometimes boring, but very important tasks serves two important functions. The first of these is that a manager can use his limited staff in other pursuits that might involve more challenging research or duties, therefore allowing him the ability to be more productive. The second function is that the person whose job has been automated now has the opportunity to be involved in more personally rewarding pursuits. Some think that the person whose function has been automated will feel threatened because he is fearful of losing his job to the robot. This tends to be a holdover from industrial robotic operations where people were displaced by robots. In fact the reverse tends to be true. Discussions with laboratory managers who have installed robots tend to indicate that their staff members now have a new challenging job, are released from routine positions, and are excited about being involved in robotics.

A related area where one can better utilize staff is in repetitive operations that require 5–10 minutes to complete. Without automation the person conducting the experiment must be with the instrument at all times since the time to do each sample is too short to allow any other tasks to be accomplished, so the experimenter's time cannot be utilized to the best advantage and he is reduced to an "instrument watcher."

Laboratory robotics also allows the removal of people from potentially hazardous environments. In *Advances in Laboratory Automation Robotics*,[8] E. Seibert of Ortho Pharmaceuticals reported the automation of the preparation of oral contraceptive samples for liquid chromatography. This automation reduced an analyst's exposure to potentially harmful steroids. Robotics lends itself to other potentially hazardous environments, such as areas where work is done with a wide variety of chemicals or laboratories involved with low-level radiation. One would envision that as the robots become more sophisticated they will be implemented in higher radiation environments.

There is also the reverse scenario: the contamination of sensitive, time-consuming experiments by humans. Some laboratory operations require an extreme amount of skill and finesse. These operations are limited to a select number of people or one person that can do the technique. A recent article about one of these techniques stated that people with such skills are guarded by their employers and could almost be traded like baseball players. If these specialized skills are amenable to automation, this potential problem could be eliminated, thereby allowing the manager additional staffing flexibility and releasing individuals to become involved in more rewarding laboratory pursuits.

Automation through the use of robotics, will also reduce the possibility of inadvertent contamination of experiments, since the possibility of human intervention will be reduced. Additionally, a robot allows a task to be accomplished repeatedly in the same exact way and in the same amount of time. For example, in biochemical reactions, timing is extremely critical and a robot could perform the same series of tasks in precisely the same way each time the series was repeated.

In addition to those items mentioned previously, there are other factors to be considered when purchasing a lab robot.[4] These include improved precision, improved sample turnaround, better data reduction and documentation, and ease of method transfer. Since a robot accomplishes each task in the same exact manner, it allows for improved precision. Each sample will have the same history and can be easily documented. Human (i.e., indiscriminate) errors are therefore eliminated. The ability to provide better data reduction and documentation is also critical. Laboratory robotic systems are equipped with a printer that allows for sample history. With proper programming, this can include sample identification, sample weight, amount of diluent, time that the analysis was started, and total time needed to accomplish the assay. This information can then be coupled with information from a variety of analytical instruments including spectrophotometers, HPLCs, GCs, and NMRs to arrive at an extremely well-documented sample. This documentation will be beneficial in several areas. It will eliminate transcription errors when data is transferred from worksheets to notebooks. Additionally, should the result be different than expected it can be rigorously checked since its history is intact and well-documented and any deviations from the experimental procedure will be clearly evident. A third benefit is the ability to provide this sample history to any individual inspecting or evaluating a particular procedure.

Another item that must also be considered, especially for a multiple plant/laboratory organization where there may be extreme distances between individual locations, is the possibility of extremely easy method transfer. In many cases method transfer between different laboratories is difficult even when one laboratory provides the other with an extremely well-documented method. Due to individual variations, certain operations are conducted in different manners or chemicals are mixed in different ways which can ultimately alter the final result. For example, a scientist in one lab specifies that a solution consisting of a mixture of three components in a certain volume ratio be made. This is written up in the method and given to another lab which attempts the procedure but fails after following all the instructions to the letter. After discussions, one of the individuals finally mentions that the order of the addition of all of the reagents is of great importance and if this is not done in an exact way the procedure will not work. If this step were automated via robotics, the laboratory that developed the method would provide the method to the other lab which had a system configured in the same way as the originating lab and the procedure would then be run in the exact same way in both labs with a very short time devoted to method transfer. The amount of time saved would be substantial and the nervousness that usually accompanies method transfer would be reduced.

Robotics will allow a laboratory to better utilize equipment on a more efficient basis. If you have spent $100,000 on analytical equipment, and you need additional capacity, you usually can't afford to buy two of these pieces of equipment, but you can use robotics to allow you to use the equipment at times when it would usually be idle. As a corollary to this premise, robotics allows a researcher to gain full usage of the current capacity rather than purchase additional units. A simplistic example is in the area of HPLC autoinjectors. If a lab needs an additional HPLC autoinjector, it would seem more reasonable to use an autoinjector associated with the robot at times when the robot wasn't active than to purchase an additional liquid chromatography (LC), autoinjector at $5000–$10,000. This allows both units full use and eliminates the purchase of unneeded equipment. These monies could then be saved and budgeted for other laboratory equipment.

This leads to another aspect of justification, the ability of a robot to accomplish multiple applications in the same configuration or with some minor changes. The current generation of automation seems basically limited to autosamplers and autoinjectors or automated instrumentation designed to accomplish a specific function. This equipment accomplishes these tasks extremely well but is limited to these specific tasks, a properly configured robot can accomplish multiple tasks allowing the lab increased productivity and better equipment utilization. Zymark Corporation[10] has placed the economic reasons for the justification of a laboratory robot into a workable form where blocks are filled in and results calculated. Samples of completed Zymark forms are included for reference.

The concept of justification of a robot will have certain generalities with all applications but each individual organization will have its own unique set of circumstances associated with its acquisition of a unit. These will have to be considered on that individual basis.

JUSTIFICATION GUIDELINES—ZYMATE LABORATORY AUTOMATION SYSTEM[11]

Economic justification of laboratory instrumentation is new. Since we're looking for very rapid recovery of the entire investment, we needn't

30   **CHAPTER 3**

make sophisticated return on investment calculations. Therefore, the justification analysis can be relatively simple.

Justification Workskeet

The enclosed justification worksheet can be used to compare any present method with an automated method using the Zymate System. Each step in the worksheet introduces a single input or calculation so that the logic behind the anslysis remains clear and simple.

An ideal project would offer full investment recovery in less than one year. With very rapid payback, the analysis can use simplifying assumptions.

To further simplify the analysis, eliminate any time or equipment which is common to both the present method and the automated method.

The following comments apply to each lettered step in the Worksheet:

A.   Number of samples per sample group
        Rather than normalize the analysis to one sample, it is convenient to define a sample group. A sample group might be a typical sample batch or the quantity run in a half or full day. Possible sample groups may be 25, 50, or 100 samples.

B.   Total time per sample group (hr)
        This is the elapsed time required to process a sample group. A technician or scientist may or may not be present.
        For estimating purposes, a robotic procedure will average about the same speed as the manual procedure.

C.   Number of sample groups per day
        This is the number of sample groups that will be run during a 24-hr period under each alternative method.

D.   Operating hours per day
        This is the time required to complete the daily number of sample groups. Under attended operation in a single shift laboratory, this time should not exceed 8 hr. If unattended operation is possible, this time may approach 24 hr.

E.   Technician cost per hour (including fringe)
        For estimating purposes use the following:
            Low cost = $12.00/phr
            Medium cost = $15.00/phr
            High cost = $20.00/phr
            Scientist = $30.00/phr

F.   Technician hours per operating day
        This is the total staffed time required for the day's work. If a person is working full time on the procedure, this will be the same as item D.

| | Operating hours per day. For automated procedures include the setup and cleanup time in this category.
G. | Technician hours per sample group
| | This is calculated to distribute the total technician hours between each sample group.
H. | Technician cost per sample group
| | The calculated labor cost for each sample group.
I. | Instrumentation cost
| | Include the purchase cost of all instrumentation and equipment required for each alternative. Common equipment can be eliminated from the comparison.
| | Certain accessories such as printers and computer interfaces provide added functionality in the automated system compared to the manual system. These may be excluded from the direct comparison.
| | For estimating purposes, typical Zymate System costs are:
| | Relatively simple system = $25,000
| | Average system = $32,000
| | Complex system = $38,000 or more
J. | Estimated user set-up and programming cost
| | The following guidelines are based on typical experience of new users installing their first Zymate System.
| | Relatively simple system: 3–5 weeks = $5,000
| | Average system: 6–10 weeks = $10,000
| | Complex system: 3–6 months = $20,000
| | Subsequent applications should require significantly less time.
K. | Total investment
| | Calculated sum of instrumentation and setup investment.
L. | Annualized investment
| | Calculated average yearly cost amortized over 5 years.
M. | Investment cost per sample group
| | Calculated daily cost over 250 working days per year. Some laboratories work over weekends and they may choose to increase the days/year to 300 or 350.
N. | Investment cost per sample group
| | Calculated cost per sample group.
O. | Total cost per sample group
| | Total cost per sample group for technician time and instrumentation.
P., Q., R. | Calculated savings and time required to recover the entire investment.

Examples

Example 1

Automatic pH titration

This is a relatively simple procedure with moderate sample quantities. Pipet the sample titrate to a pH end point.

Conclusion

By running unattended for 16 hr, they were able to complete five sample batches each week and recover their investment in about nine months.

	Formula	Present method	ZYMATE SYSTEM alternative 1
A. Number of samples per sample group	Input	100	100
B. Total time per sample group (hr)	Input	16.0	16.0
C. Number of sample groups per day	Input	0.5	1.0
D. Operating hours per day	B×C	8.0	16.0
E. Technician cost per hour (including fringe)	Input	$12.00	$12.00
F. Technician hours per operating day	Input	8.0	0.0
G. Technician hours per sample group	F/C	16.0	0.0
H. Technician cost per sample group	E+G	$192.00	$0.0
I. Instrumentation cost	Input	—	$25,000
J. Estimated user setup and programming cost	Input	—	$5,000
K. Total investment	I+J	$0	$30,000
L. Annualized investment	K/5	$0	$6,000
M. Investment cost per day	L/250	$0.00	$24.00
N. Investment cost per sample group	M/C	$0.00	$24.00
O. Total cost per sample group	H+N	$192.00	$24.00
P. Zymate system saving per sample group=0. (present method)−0. (Zymate System method)	—	—	$168.00
Q. Number of sample groups to recover investment	K/P	—	178.6
R. Days to recover investment	Q/C	—	178.6

Typical automated Karl Fischer payback justification

The following payback analysis assumes that automation will permit unattended 24 hr/day operation and that the sample throughput rate is the same for both the manual and automated procedure. The cost of the titrator has been left out of the analysis because it is required for both the manual and automatic procedures.

THE JUSTIFICATION OF A LABORATORY ROBOT

	Formula	Manual use of titrator	Fully automated titration system
A. Number of samples per sample group	Input	30	30
B. Total time per sample group (hr)	Input	4	4
C. Number of sample groups per day	Input	2	6
D. Operating hours per day	B×C	8	24
E. Technician cost per hour (including fringe)	Input	$15	$15
F. Technicial hours per operating day	Input	8	3
G. Technician hours per sample group	F/C	4	0.5
H. Technician cost per sample group	E+G	$60	$7.50
I. Instrumentation cost	Input	—	$40,000
J. Estimated user setup and programming cost	Input	—	$10,000
K. Total investment	I+J	—	$50,000
L. Annualized investment	K/5	—	$10,000
M. Ownership cost per day	L/250	—	$40
N. Ownership cost per sample group	M/C	—	$6.67
O. Total cost per sample group	H+N	$60	$14
P. Zymate System saving per sample group=0. (present method)−0. (Zymate System method)	—	—	$46
Q. Number of sample groups to recover investment	K/P	—	
R. Days to recover investment	Q/C	—	181

Example 2

Typical research or methods development applilcation.

Labor-intensive, moderately difficult procedure.

Zymate System alternative 1 permits 2 sample groups per day and alternative 2 permits 3 sample groups per day.

Instrumentation needed for manual procedures includes such items as semiautomatic dispensers, electronic balances, centrifuges, and shakers.
Conclusions

Alternative 1 generates a payback in about one year and alternative 2 in about 6 months.

34 CHAPTER 3

	Formula	Present method	ZYMATE SYSTEM alternative 1	ZYMATE SYSTEM alternative 2
A. Number of samples per sample group	Input	20	20	20
B. Total time per sample group (hr)	Input	8.0	8.0	8.0
C. Number of sample groups per day	Input	1.0	2.0	3.0
D. Operating hours per day	B+C	8.0	16.0	24.0
E. Technician cost per hour (including fringe)	Input	$15.00	$15.00	$15.00
F. Technician hours per operating day	Input	8.0	3.0	3.0
G. Technician hours per sample group	F/C	8.0	1.5	1.0
H. Technician cost per sample group	E+G	$120.00	$22.50	15.00
I. Instrumentation cost	Input	$4,000	$32,000	$32,000
J. Estimated user setup and programming cost	Input	$1,000	$10,000	$10,000
K. Total investment	I+J	$5,000	$42,000	$42,000
L. Annualized investment	K/5	$1,000	$8,400	$8,400
M. Investment cost per day	L/250	$4.00	$33.60	$33.60
N. Investment cost per sample group	M/C	$4.00	$16.80	$11.20
O. Total cost per sample group	H+N	$124.00	$39.30	$26.20
P. Zymate System saving per sample group 0. (present method)−0. (Zymate System method)	—	—	$84.70	97.80
Q. Number of sample groups to recover investment	K/P	—	495.9	429.4
R. Days to recover investment	Q/C	—	247.9	143.1

Example 3

High volume, repetitive.

Relatively simple procedure with large sample quantity requirements.

An optimized Zymate System layout and program permits the automated system to run 10% faster than the usual procedure.

Conclusion

By running 6 sample groups or 600 samples unattended over a 21.6 hr period, the investment is recovered in about 9 months.

		Formula	Present method	ZYMATE SYSTEM alternative 1
A.	Number of samples per sample group	Input	100	100
B.	Total time per sample group (hr)	Input	4.0	3.6
C.	Number of sample groups per day	Input	2.0	6.0
D.	Operating hours per day	B+C	8.0	21.6
E.	Technician cost per hour (including fringe)	Input	$12.00	$12.00
F.	Technician hours per operating day	Input	8.0	6.0
G.	Technician hours per sample group	F/C	4.0	1.0
H.	Technician cost per sample group	E+G	$48.00	$12.00
I.	Instrumentation cost	Input	$2,500	$32,000
J.	Estimated user setup and programming cost	Input	$500	$5,000
K.	Total investment	I+J	$3,000	$37,000
L.	Annualized investment	K/5	$600	$7,400
M.	Investment cost per day	L/250	$2.40	$29.60
N.	Investment cost per sample group	M/C	$1.20	$4.93
O.	Total cost per sample group	H+N	$49.20	$16.93
P.	Zymate System saving per sample group=0. (present method)−0. (Zymate System Method)		—	—	$32.27
Q.	Number of sample groups to recover investment	K/P	—	1146.9
R.	Days to recover investment	Q/C	—	191.1

Examples 4 and 5

Sample introduction to expensive instrumentation.

Expensive instrumentation such as NMR spectrometers and physical testing equipment require people for sample introduction. In most laboratories, their use is limited to the normal 8 hr workday.

Example 4 illustrates the benefit of adding automation to an existing instrument and Example 5 illustrates the benefit of automating one instrument for extended use rather than purchasing a second instrument.
Conclusions

In Example 4 automating during the 8 hr normal workday has limited economic benefit, but with extended use the investment can be recovered in about 1 year.

	Formula	Present method[a]	ZYMATE SYSTEM alternative 1[b]	ZYMATE SYSTEM alternative 2[c]
A. Number of samples per sample group	Input	30	30	30
B. Total time per sample group (hr)	Input	7.0	7.0	7.0
C. Number of sample groups per day	Input	1.0	1.0	2.0
D. Operating hours per day	B+C	7.0	7.0	14.0
E. Technician cost per hour (including fringe)	Input	$15.00	$15.00	$15.00
F. Technician hours per operating day	Input	7.0	1.5	3.0
G. Technician hours per sample group	F/C	7.0	1.5	1.5
H. Technician cost per sample group	E+G	$105.00	$22.50	$22.50
I. Instrumentation cost	Input	$0	$25,000	$25,000
J. Estimated user setup and programming cost	Input	$0	$10,000	$10,000
K. Total investment	I+J	$0	$35,000	$35,000
L. Annualized investment	K/5	$0	$7,000	$7,000
M. Investment cost per day	L/250	$0.0	$28.00	$28.00
N. Investment cost per sample group	M/C	$0.0	$28.00	$14.00

THE JUSTIFICATION OF A LABORATORY ROBOT

		Formula			
O.	Total cost per sample group	H+N	$105.00	$50.50	$36.50
P.	Zymate System saving per sample group=0. (present method)−0. (Zymate System method)		—	$54.50	$68.50
Q.	Number of sample groups to recover investment	K/P	—	642.2	510.9
R.	Days to recover investment	Q/C	—	642.2	255.5

[a] Present method—manual sample introduction: daytime only.
[b] Zymate System alternative 1—automatic sample introduction: daytime only.
[c] Zymate System alternative 2—automatic sample introduction: day plus evening.

In Example 5 the first instrument is automated for extended use, compared to purchasing a second instrument. In this case the payback is less than 6 months.

		Formula	Present method[a]	ZYMATE SYSTEM alternative 1[b]
A.	Number of samples per sample group	Input	30	30
B.	Total time per sample group (hr)	Input	7.0	7.0
C.	Number of sample groups per day	Input	1.0	2.0
D.	Operating hours per day	B+C	7.0	14.0
E.	Technician cost per hour (including fringe)	Input	$15.00	$15.00
F.	Technician hours per operating day	Input	7.0	3.0
G.	Technician hours per sample group	F/C	7.0	1.5
H.	Technician cost per sample group	E+G	$105.00	$22.50
I.	Instrumentation cost	Input	$100,000	$25,000
J.	Estimated user setup and programming cost	Input	$0	$10,000
K.	Total investment	I+J	$100,000	$35,000

L.	Annualized investment	K/5	$20,000	$7,000
M.	Investment cost per day	L/250	$80.00	$28.00
N.	Investment cost per sample group	M/C	$80.00	$14.00
O.	Total cost per sample group	H+N	$185.00	$36.50
P.	Zymate System saving per sample group 0. (present method)−0. (Zymate System method)	—	—	$148.50
Q.	Number of sample groups to recover investment	K/P	—	235.7
R.	Days to recover investment	Q/C	—	117.8

[a]Present method—manual sample introduction: daytime only.
[b]Zymate system alternative 1—automatic sample introduction: daytime only.
[c]Zymate system alternative 2—automatic sample introduction: day plus evening.

Additional Justification

Comment on any noneconomic benefit which contributes to justifying the Zymate Laboratory System.

1. Improved precision—

2. Faster sample turnaround time—

3. Automatic data reduction and documentation—

4. Safety—

5. New methods—

6. Other applications—

4
Applications for Laboratory Robotics

Laboratory robotics provides flexible automation required to meet the changing needs typical of industrial and research laboratories. Flexible automation is programmed by the user to perform multiple procedures. As explained in the preceding chapters, it is not necessary to have a large quantity of identical, repetitive operations to justify the purchase of a laboratory robot. Most of the procedures found in the laboratory today have the potential to be automated using robotics.

An automation system consists of a robot, a controller, laboratory stations, and various laboratory kits. It also combines several technologies including computer science, analytical instrumentation, chemistry and robotics.

Some of the commercially available laboratory stations (Figs. 4-1 through 4-12) used in various applications include the following:

1. **Power and event control station.** Useful in bringing most standard laboratory apparatus under system control. It provides programmable control of on/off and variable electric power and switch closures. It also senses external switch closures and analog voltages.
2. **Master laboratory station.** Provides liquid handling, extraction, and partition capability. It utilizes three computer-controlled syringes that can be programmed to dispense, dilute, and pipet.
3. **Instrument interface.** Provides programmable data acquisition from laboratory instruments and programmable control of other apparatus in the laboratory.

CHAPTER 4

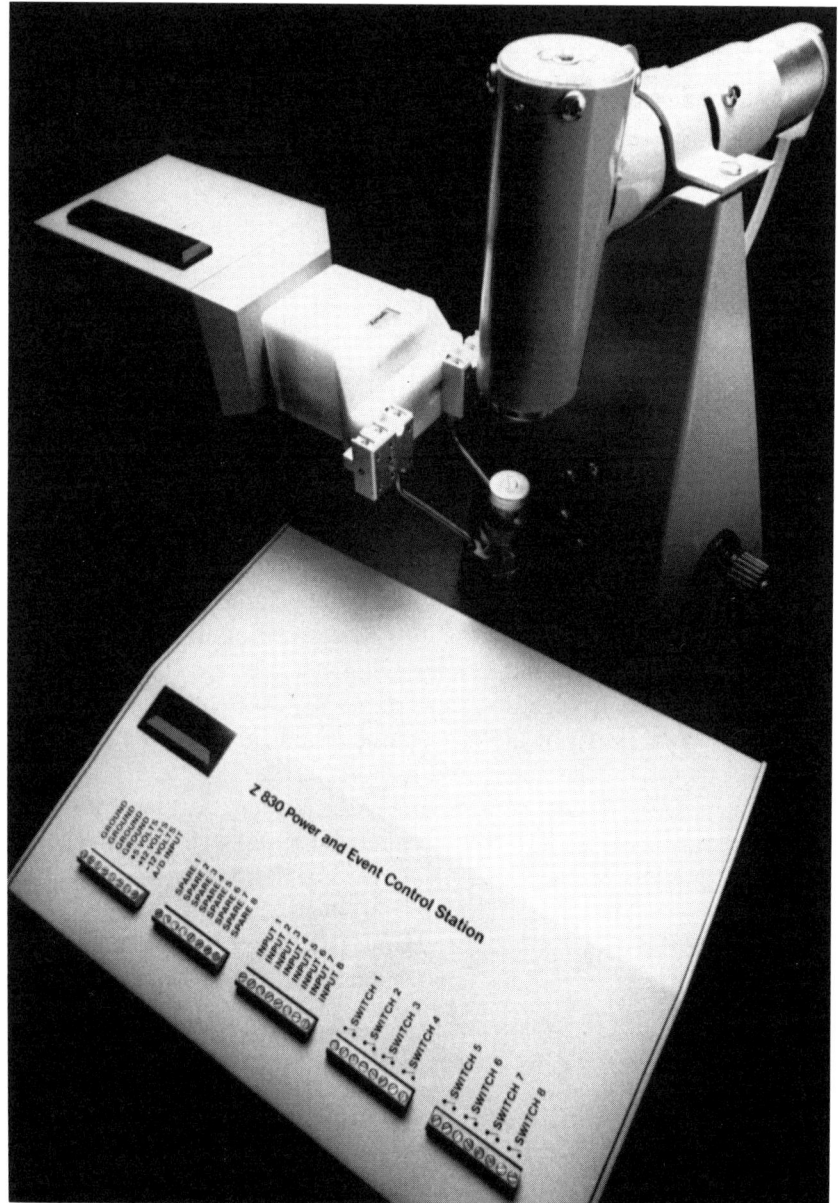

FIGURE 4-1. Power and event control station. (Photo courtesy of Zymark Corporation.)

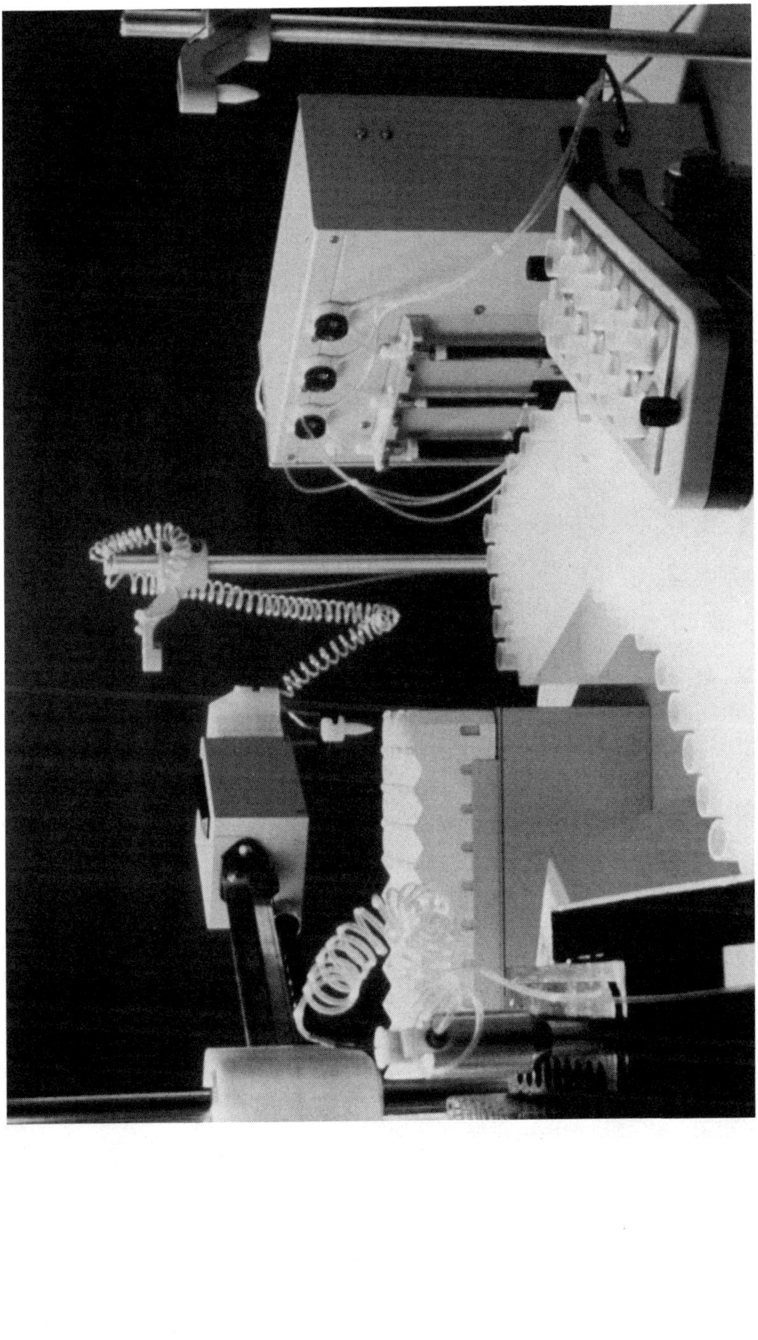

APPLICATIONS FOR LABORATORY ROBOTICS

FIGURE 4-2. Master laboratory station. (Photo courtesy of Zymark Corporation.)

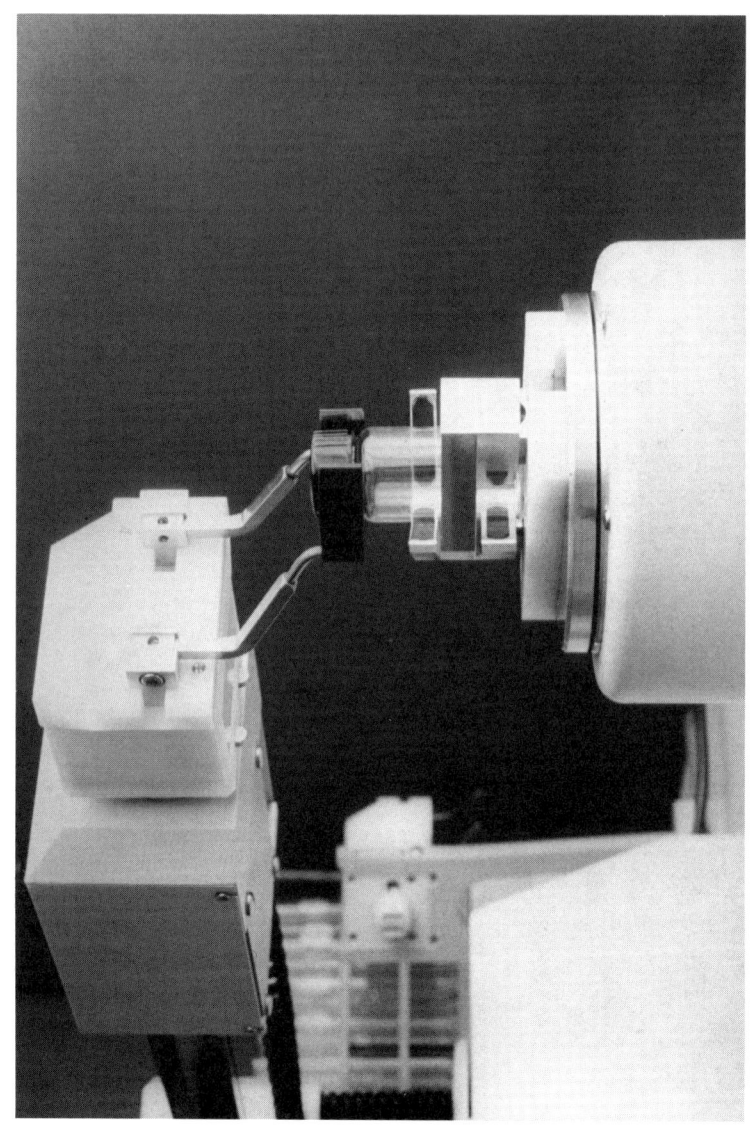

FIGURE 4-3. Robot capping station. (Photo courtesy of Zymark Corporation.)

APPLICATIONS FOR LABORATORY ROBOTICS 43

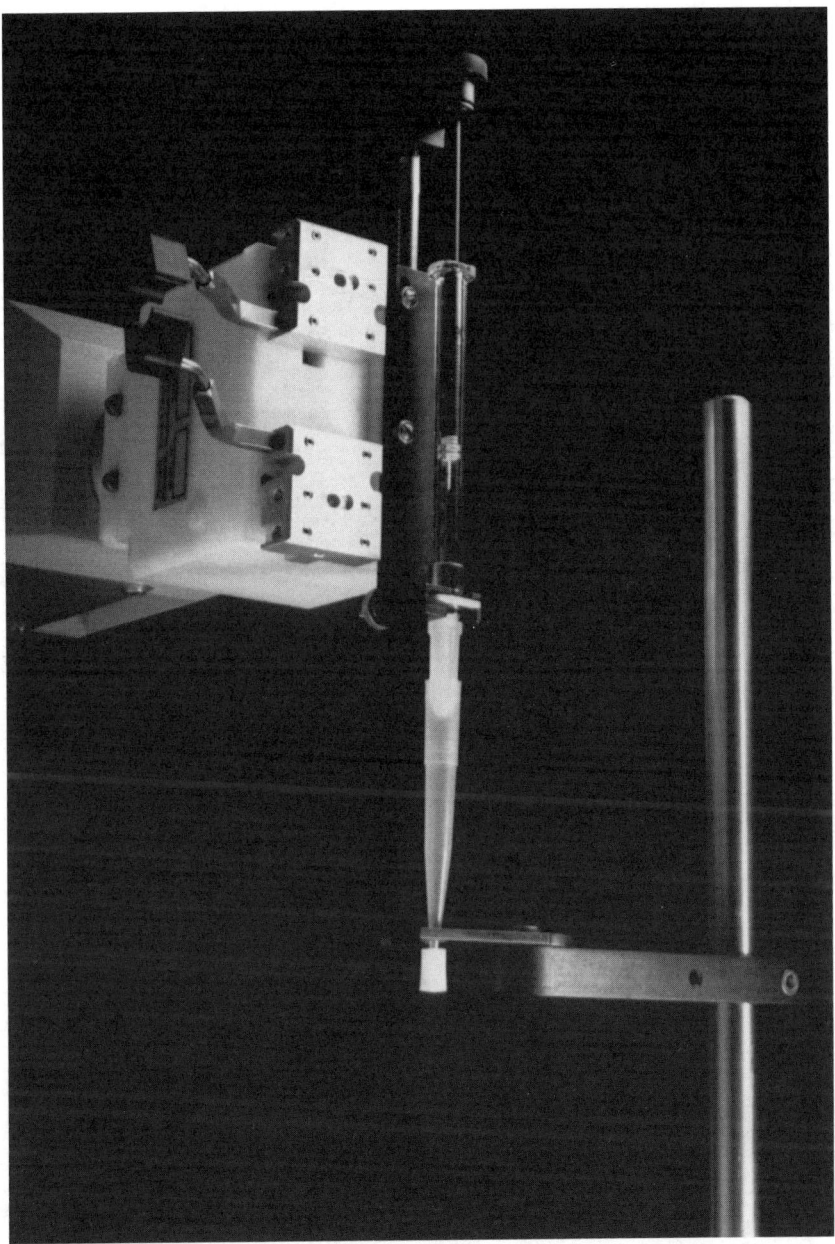

FIGURE 4-4. Syringe hand. (Photo courtesy of Zymark Corporation.)

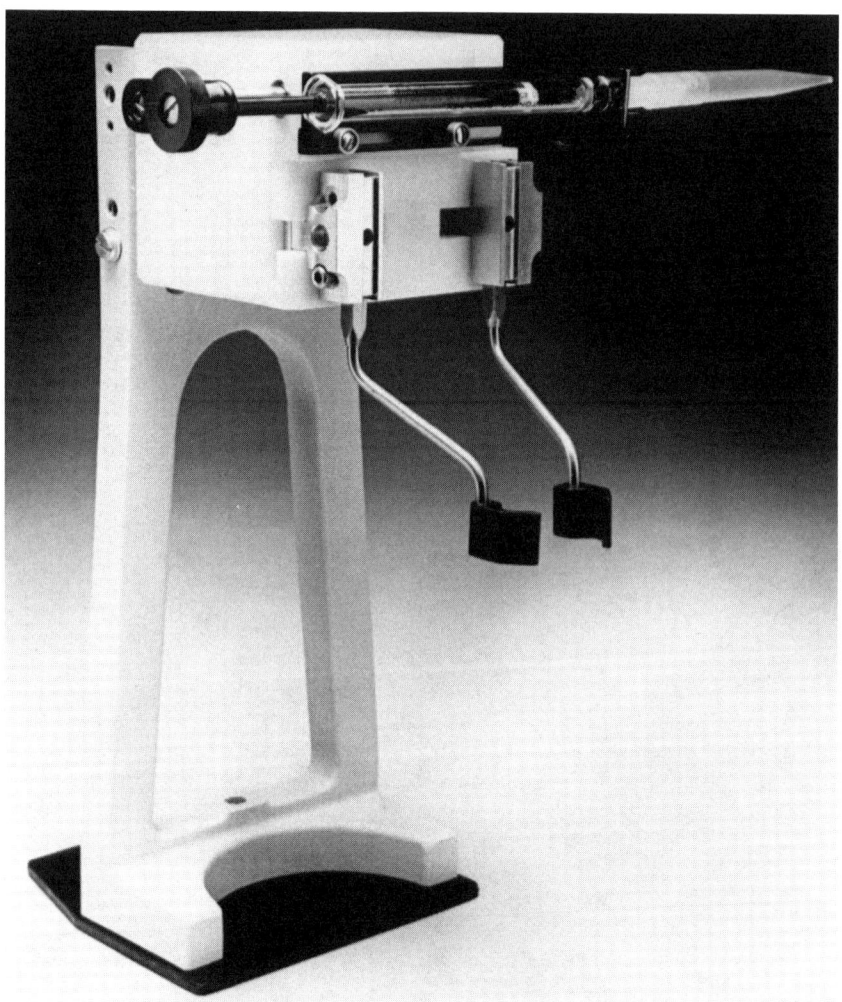

FIGURE 4-5. Dual function hand. (Photo courtesy of Zymark Corporation.)

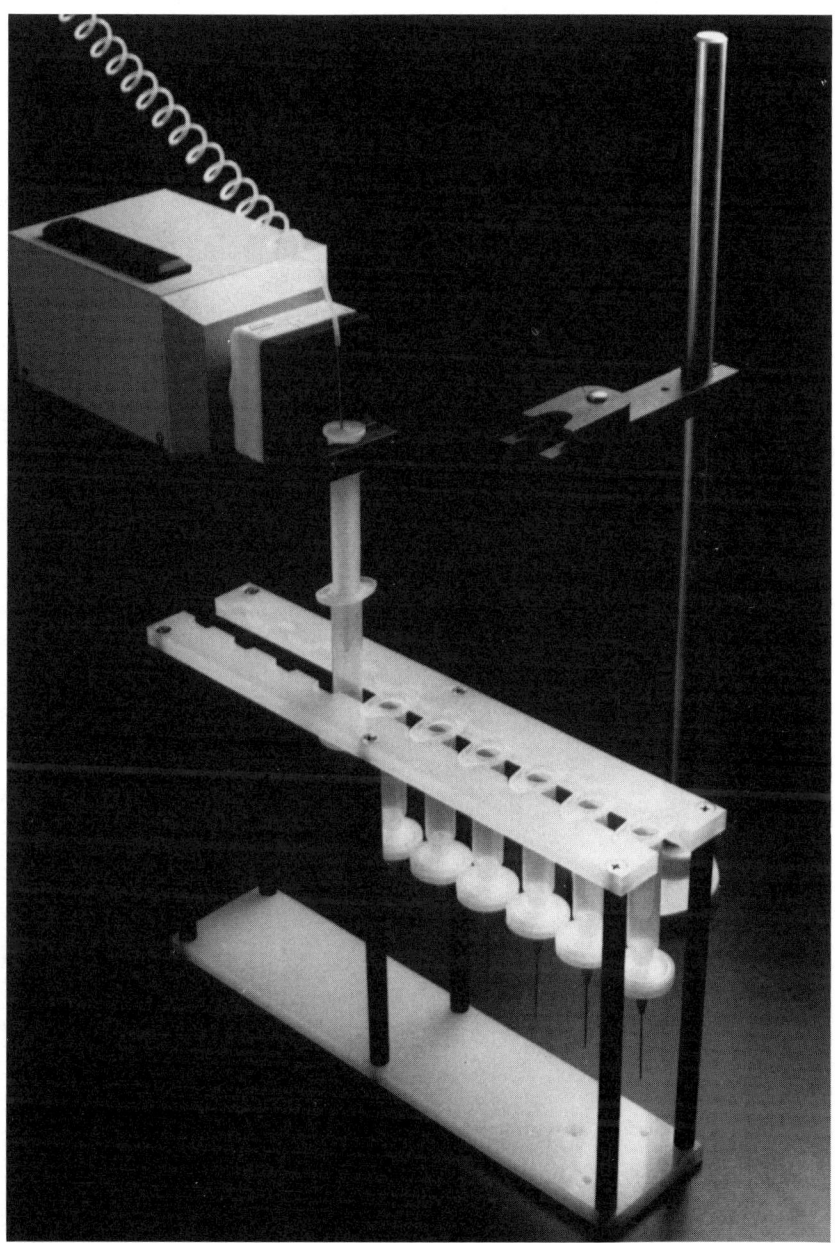

FIGURE 4-6. Liquid distribution hand. (Photo courtesy of Zymark Corporation.)

FIGURE 4-7. Robot used for automated ICP analysis of laboratory oils (Photo courtesy of "Advances in Laboratory Automation—Robotics," G. L. Hawk and J. R. Strimaitis, 1984.)

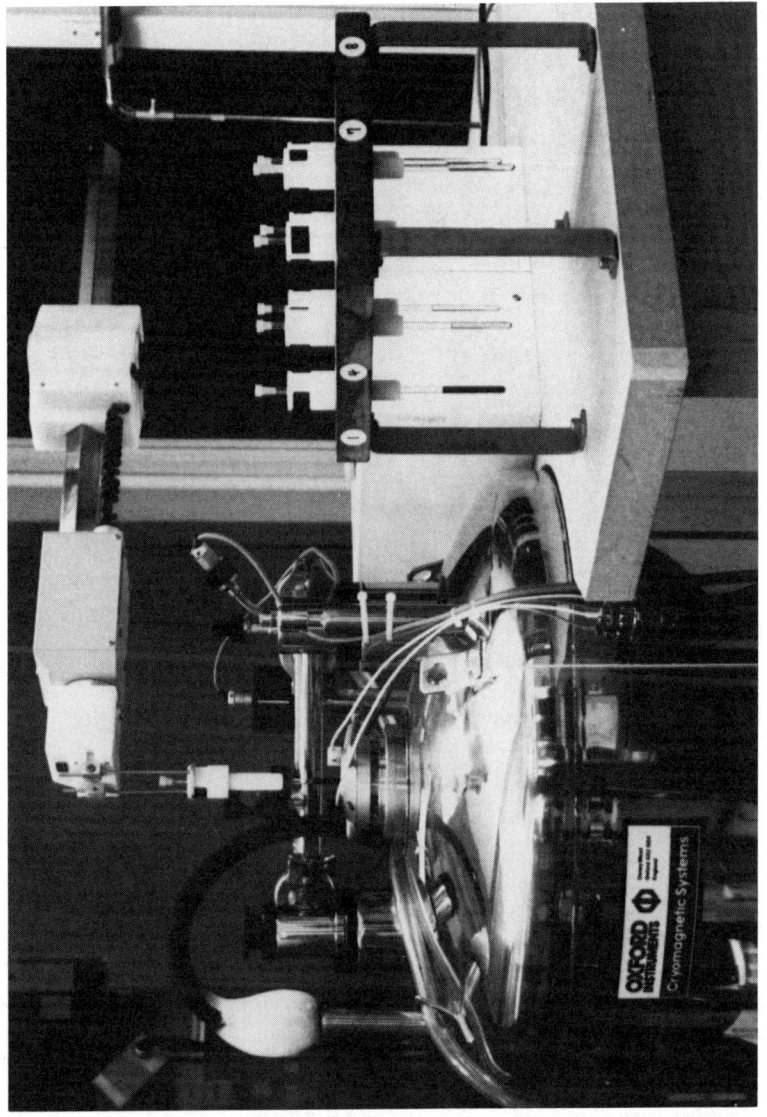

FIGURE 4-8. Robot loading superconducting NMR (Photo courtesy of "Advances in Laboratory Automation—Robotics," G. L. Hawk and J. R. Strimaitis, 1984.)

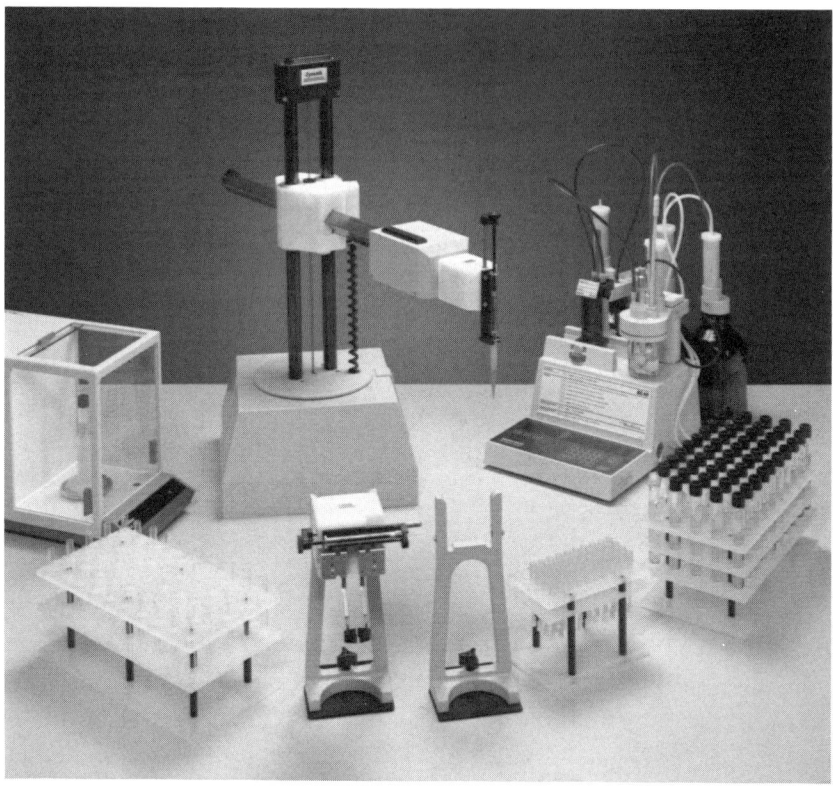

FIGURE 4-9. Robot configuration for automatic Karl Fischer moisture determination. (Photo courtesy of Zymark Corporation.)

4. **Capping Station.** Laboratory station for capping round containers with screw caps. This station grasps and turns a wide range of container sizes and coordinates its motion with that of the robot and a general purpose hand (GP hand).
5. **General Purpose gripper hand (GP hand).** A hand that provides a basic capability to grasp and move vials, test tubes, beakers, and other laboratory devices.
6. **Syringe hand.** A hand that has a motor-driven syringe attached. It will provide the accurate transfer of samples using a cannula or a disposable pipet tip.
7. **Dual function hand.** Combines the capability of the GP hand and the syringe hand.
8. **Powder dispensing hand.** This is a general purpose hand which includes a small motor with an off-axis weight which creates vibration in the hand. The speed of vibration can be controlled through the program and will receive feedback from a balance to hit a target weight.

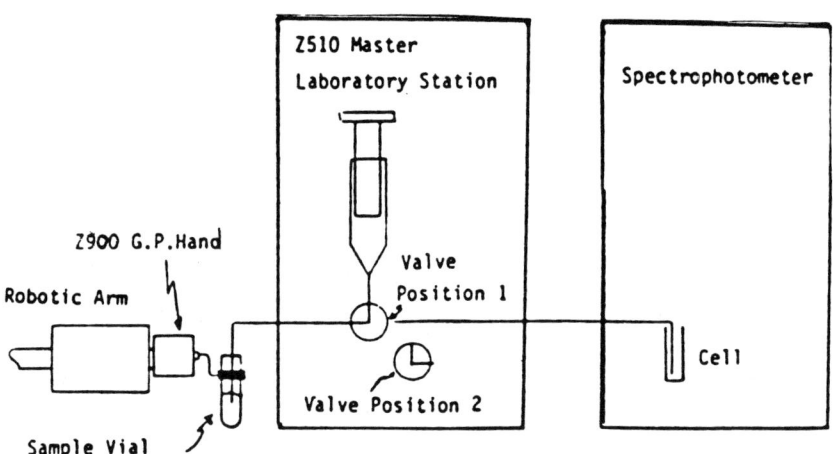

FIGURE 4-10. Automated spectroscopy using robotics. (Photo courtesy of Zymark Corporation.)

9. **Liquid distribution hand.** Used for pipeting, manifolding, and remote distribution of liquids from the master laboratory station.

These nine items are some of the commercially available laboratory stations, but are by no means the only available items. Many researchers have developed or modified other types of stations for use in a specific application. This text has chosen to concentrate on available systems and stations rather than discuss the fabrication of items to make them "robot friendly."

APPLICATIONS

Automatic Titrations

The robotic system, in combination with a pH or selective ion meter, can perform automatic end-point titrations. The analog output from the pH meter is connected to the analog to the digital converter on the power and event control station. This station allows input signals to be read by the robotic system and scaled to a particular specified range. The titrant is delivered through the nozzle of the master laboratory station. The robot moves to a rack an pick up a vial containing a sample. It then brings the vial to the pH electrode and measures the initial pH. Titrant

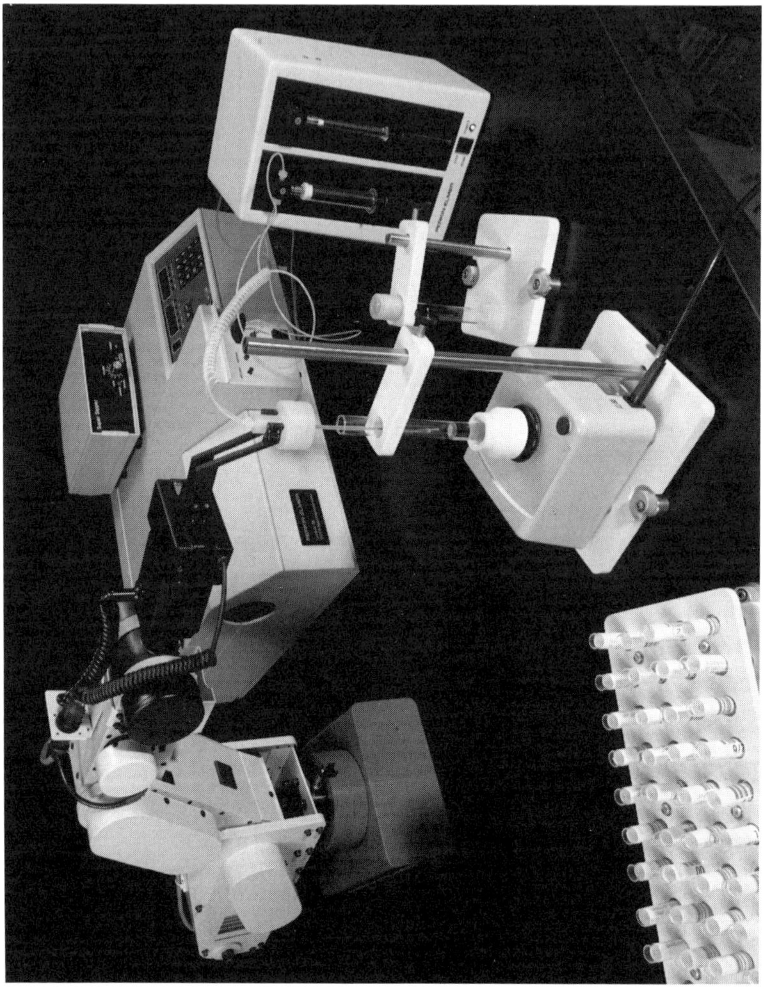

FIGURE 4-11. Robotics applied to UV/VIS Spectroscopy. (Photo courtesy of Perkin-Elmer Corporation.)

APPLICATIONS FOR LABORATORY ROBOTICS 51

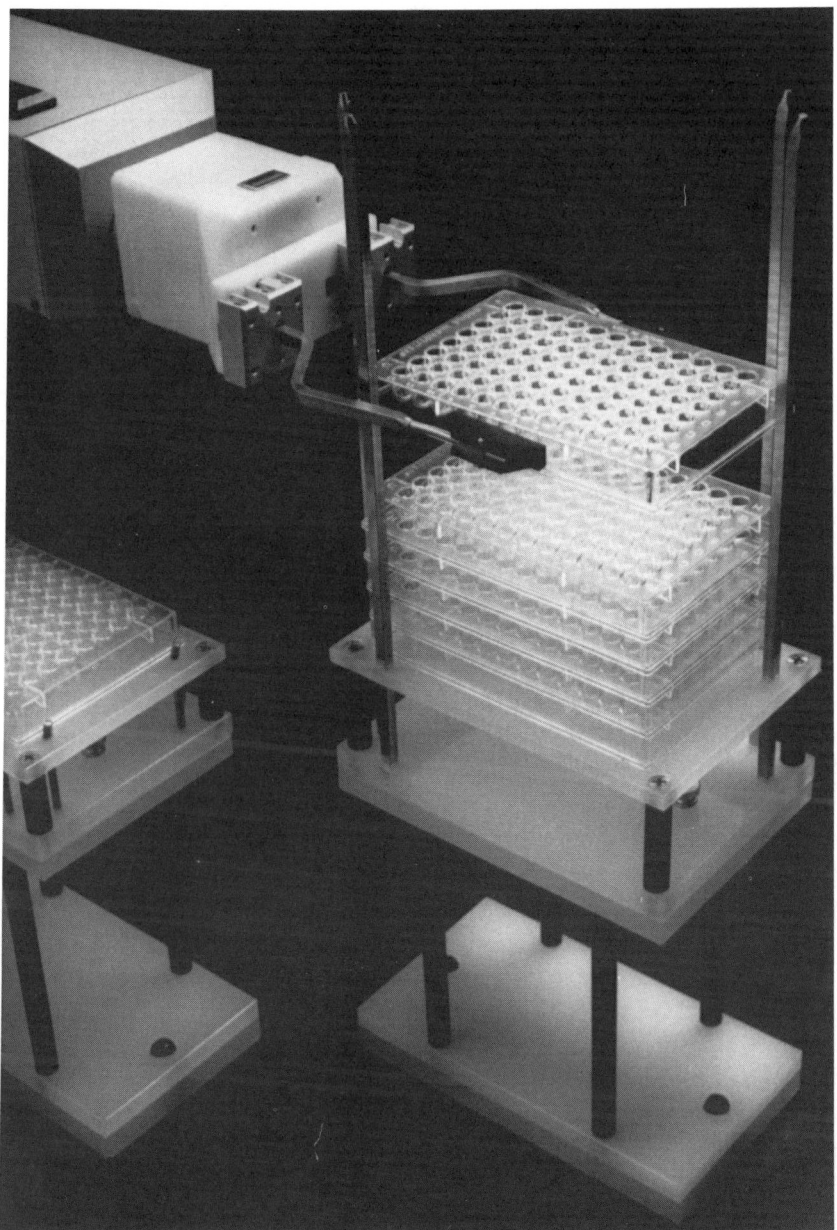

FIGURE 4-12. ELISA plates with modified general purpose hand. (Photo courtesy of Zymark Corporation.)

is then dispensed in programmed increments into the sample vial and continues to be dispensed in decreasing amounts until the desired endpoint is reached. The robotic controller calculates the total volume of titrant dispensed and the final pH. This information is then sent to a printer or a host computer.

Total Solids Determination

The robot, using a general purpose hand moves to a rack of aluminum weighing tins. The robot picks up one of the tins and places it on the pan of an electronic balance. Through the use of a **balance interface**, the weight of the tin is tared. The robot parks the general purpose hand, and attaches a syringe hand that has been modified to pick up disposable pipet tips. The robot moves to a rack of pipet tips and attaches a tip to the syringe hand. The robot moves to the sample and draws the required volume into the pipet tip. The sample is then dispensed onto the sample pan on the balance, and the sample weight is recorded in the controller. The pipet tip is automatically removed and the general purpose hand reattached. The robot removes the aluminum tin from the balance and places it into an oven for a desired period of time. Some modification will probably need to be done to the oven door to allow it to easily open and close in an automation setup. One approach is to use an air cylinder connected to the door of the oven and the frame. The power and event control station can send a signal to a valve to activate the air cylinder to push the door open and pull it closed at the needed time. When the incubation time is completed the door is signaled to open, and the robot removes the sample tin and places it in a moisture-free environment. Desiccators were not designed for robots. Therefore one possibility, alternative is to use a Plexiglas box, filled with a drying agent and fitted with a sliding door on the top that could be motor driven, opened with the air cylinder, or pushed open by the robot. After the sample is cooled, the robot places it back on the balance to be reweighed. The final weight is subtracted from the initial weight and the total solids number is calculated. This value can be printed out or sent on to a host computer.

Automated Biological Oxygen Demand (BOD) Determination

Biological oxygen demand determinations are routinely required for environmental monitoring. These repetitive procedures are well suited to

APPLICATIONS FOR LABORATORY ROBOTICS 51

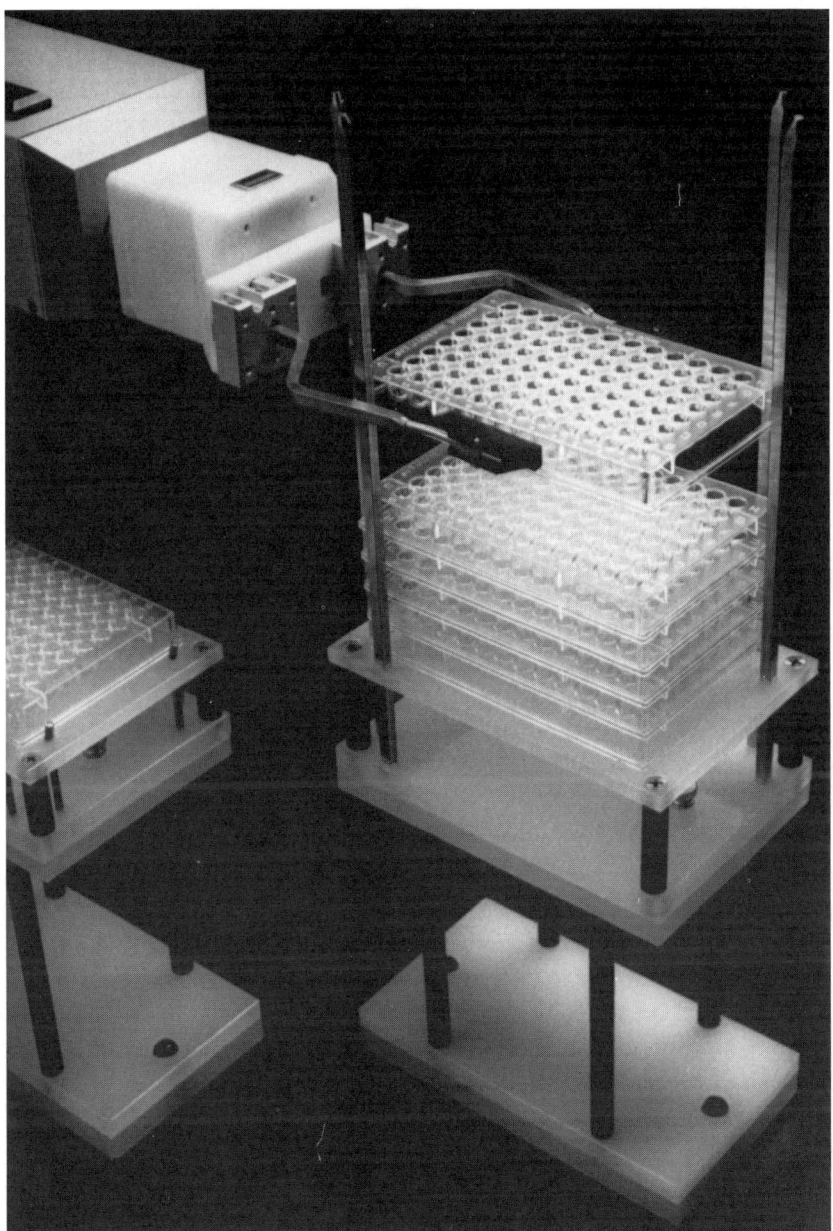

FIGURE 4-12. ELISA plates with modified general purpose hand. (Photo courtesy of Zymark Corporation.)

is then dispensed in programmed increments into the sample vial and continues to be dispensed in decreasing amounts until the desired endpoint is reached. The robotic controller calculates the total volume of titrant dispensed and the final pH. This information is then sent to a printer or a host computer.

Total Solids Determination

The robot, using a general purpose hand moves to a rack of aluminum weighing tins. The robot picks up one of the tins and places it on the pan of an electronic balance. Through the use of a **balance interface,** the weight of the tin is tared. The robot parks the general purpose hand, and attaches a syringe hand that has been modified to pick up disposable pipet tips. The robot moves to a rack of pipet tips and attaches a tip to the syringe hand. The robot moves to the sample and draws the required volume into the pipet tip. The sample is then dispensed onto the sample pan on the balance, and the sample weight is recorded in the controller. The pipet tip is automatically removed and the general purpose hand reattached. The robot removes the aluminum tin from the balance and places it into an oven for a desired period of time. Some modification will probably need to be done to the oven door to allow it to easily open and close in an automation setup. One approach is to use an air cylinder connected to the door of the oven and the frame. The power and event control station can send a signal to a valve to activate the air cylinder to push the door open and pull it closed at the needed time. When the incubation time is completed the door is signaled to open, and the robot removes the sample tin and places it in a moisture-free environment. Desiccators were not designed for robots. Therefore one possibility, alternative is to use a Plexiglas box, filled with a drying agent and fitted with a sliding door on the top that could be motor driven, opened with the air cylinder, or pushed open by the robot. After the sample is cooled, the robot places it back on the balance to be reweighed. The final weight is subtracted from the initial weight and the total solids number is calculated. This value can be printed out or sent on to a host computer.

Automated Biological Oxygen Demand (BOD) Determination

Biological oxygen demand determinations are routinely required for environmental monitoring. These repetitive procedures are well suited to automation using laboratory robotics. The method described is based on dissolved oxygen electrode measurements prior to and following a 5-day incubation.

The first step is to adjust the sample pH to 7.0 The robot picks up the sample container and brings it to a mounted pH probe that has been connected to the analog to digital converter on the power and event controller, and makes an initial measurement. The controller calculates the amount of buffer to be dispensed by the master laboratory station to reach a pH of 7.0 based upon the measured pH. After the buffer is dispensed the robot transfers the sample to a standard BOD bottle by a pouring, and then fills it with the diluent using a metering pump controlled by the power and event controller. The dissolved oxygen (DO) is measured by having the robot bring the DO probe to the sample. The DO meter is connected to the analog to digital converter on the power and event controller. The DO value is stored in the controller or transmitted to a laboratory computer. The robot caps the BOD bottle which is moved manually to an incubator where it is incubated for 5 days. After incubation, the samples are returned to the robotic work area. The robot uncaps the BOD bottle and measures the dissolved oxygen. The results are printed or sent on to another computer for permanent documentation.

Sample Preparation of Polymer Samples for Gel Permeation Chromatography (GPC)

The robot moves to a rack of clean screw-cap vials. It brings the vial to the capping station where it removes the cap and places it temporarily in a holding area. The vial is then placed on the balance and tared. The next step is to weigh out the desired amount of sample, a procedure to which there are several approaches. The robot could attach a power dispensing hand, a specially modified general purpose hand that allows the robot to reliably transfer powders and other solid samples from one container to another. This hand is equipped with a program-controlled, variable-speed electric motor on which is mounted an eccentric mass. The robot picks up the sample container and places it above the receiving container on the balance at a pouring angle of perhaps 90 to 100 degrees, then the vibration starts, continuing until the sample target weight is reached. Because there are many parameters that effect the rate and efficiency of powder pouring it is not possible to give specific directions. In setting up a program to pour a solid, some experimentation will have to be done to find the right combination of pouring angle and vibration intensity for a particular sample. Another approach to weighing is to have a representative amount of sample manually placed in a container and have the robot pour the contents into the receiving vial on the balance. The actual weight would then be calculated and the volume of solvent to be added later would be adjusted to hit a specific target concentration. The final approach to weighing is to have the robot tare a number of empty vials and place them back into their rack. A representative amount of sample would then be manually added to the container. The robot would then reweigh the sample vial and calculate the actual sample weight and adjust the solvent required to achieve the desired concentration. It would be valuable here to have an "If....Then" statement in your program to reject samples that fall outside the tolerable weight limits.

Using an accurate variable volume dispenser such as the master laboratory station, the required volume of solvent is added to reach the desired concentration. The robot then brings the sample back to the capping station, retrieves the original cap from its holding area, and recaps the vial. The next step is to mix the sample. The robot places the sample in a rack that is fitted on a linear shaker. The shaker must be modified to start and stop in the same location each time so the robot can reliably find the samples it places in the rack. The controller keeps track of the length of time each sample shakes so that all samples have the same history of preparation. When the mixing is completed the sample needs to be filtered. Most GPC analysis requires filtration through a 0.45 micron filter. A popular approach has been to have rack of disposable syringe barrels, each fitted with a $0.45v$ filter. After uncapping the sample, the robot attaches a syringe hand and picks up a disposable pipet tip and draws the required amount of sample, then dispenses it into the syringe barrel/filter setup. The pipet tip is automatically removed, the syringe hand is parked, and a blank hand fitted with a remote nozzle in the barrel is attached, and the sample is forced through the filter by air pressure coming from the master lab station or other controlled air supply. The filtered sample is then collected in an autosampler vial. The robot then caps the vial and places it into a carousel for an autoinjector or makes the direct injection into the GPC.

Automated Differential Scanning Calorimetry (DSC)

The robot can be considered the universal autosampler. This capability is especially valuable for instruments that do not have any means to automatically introduce samples. The robot has been used in unattended operation to load and unload samples from multiple DSCs.

Several DSC cells are placed around the robot. It keeps track of when each sample is loaded and analysis is started. Powdered samples should be prepared in uniform sealed pans instead of the crimp-type aluminum pans. The sample pans are loaded manually into a sample rack. The robot uses a special multifunction hand, which consists of large fingers to remove the DSC bell jars and a vacuum "finger" to pick up the sample pan. The robot moves to the sample rack and picks up the first sample using the vacuum "finger" portion of the multifunction hand. The pickup is verified by depressing a microswitch located on the sample rack. The microswitch is connected to the input switch found on the power and event controller. If the microswitch closure is detected, the robot is instructed to continue with the procedure, however, if no switch closure is detected, the robot is instructed to find the next sample. The sample is transferred to the proper position in the DSC sample cell. Using the large fingers on the multifunction hand, the robot replaces the proper covers for the DSC cell. The number and type of covers can vary depending on the type of DSC application. The robot controller, through the use of a **computer interface**, starts the analysis on the DSC. The robot then starts another DSC cell if this is desired. At the completion of DSC analysis the results are printed and transmitted to a laboratory computer.

Sample Preparation for Polychlorinated Biphenyls (PCB) Analysis by Gas Chromatography (GC)

PCB content in transformer oils, hydraulic fluids, and a variety of waste oils is a vital environmental concern. PCB analysis, therefore, is widely used throughout the chemical industry to identify potentially toxic materials.

An approach to automating this procedure using the robot is as follows. The robot picks up an empty test tube, places it on the balance and tares it. Using the syringe hand, the robot attaches a disposable pipet tip and draws up the desired amount of sample. The sample aliquot is then dispensed into the tared test tube on the balance. Through the balance interface to the controller, the actual sample weight is calculated. Based upon the sample weight, an extraction solvent is added by the master laboratory station. The robot places the test tube into the vortex for one min. The vortex is controlled by the variable AC outlet on the power and event controller. After vortexing, the tube is returned to the rack. The solvent layer containing the PCB extract is withdrawn by using a movable nozzle with a 6-inch cannula that is attached to one of the syringes on the master laboratory station through coiled tubing. The extract is dispensed into a GC vial. The cannula on the nozzle is placed into a wash station to clean both the outside and the inside with an appropriate solvent. The GC vial is crimped by having the robot bring to the vial a septum cap and place it loosely on the vial. The robot now brings the vial to an air-actuated crimper. The crimper is turned on and off through the power and event controller. After crimping, the sample is placed into an autosampler carousel or loaded into a single shot autoinjector and immediately analyzed. As this sample is being run, subsequent samples can be prepared.

Thin Layer Chromatography (TLC)

Most TLC procedures can be broken into four major steps as follows:

1. TLC sample preparation
2. Spotting or streaking of plates
3. Developing plates
4. Reading plates

Sample preparation for TLC involves all of the same issues as for GC and HPLC analysis and is quite straightforward for robotic automation.

The procedure that has been used to spot a TLC plate using a laboratory robotic system is as follows:

1. Manually place a stack of TLC plates upside down in a holder. The TLC plate box is usually used as the holder and is simply placed into a locator rack.
2. The robot attaches a specially modified general purpose hand, that has been fitted with a suction attachment. Using this hand, the robot picks up a TLC plate and locates it in a "spotting station."
3. Next the robot attaches the syringe hand, takes an aliquot of sample, and spots the plate. Multiple spots and respotting can be easily performed.
4. The robot reattaches the modified general purpose hand, retrieves the plate from the spotting station, and places it in a "spotted plate holding station."

The plates are currently being transferred manually to a developing tank; this step also could be automated with the robot.

Automating Liquid/Solid Extraction

The laboratory robotic system has been used to automate many types of liquid/solid extractions. One popular technique employs the use of disposable syringe barrels packed with various types of bonded phases. Methods have been developed using robotics to interface with these columns which provides a maximum amount of flexibility, and reproducibility, as well as the ability to perform the steps before and after the extraction.

The following is a typical example to illustrate the point.

1. Prepare the sample: The robot attaches the liquid distribution hand fitted with a tapered remote nozzle to pick up disposable pipet tips as well as extraction columns. It picks up a pipet tip and transfers 1 ml of sample to a test tube (the remote nozzle on the LDH is connected to the syringes on the master lab station). It also adds 0.1 ml of an internal standard solution and dilutes to 10 ml using the master lab station. The tube is transferred to the **vortex station** where the prepared sample is mixed.
2. Prewet the column: Reverse-phase columns usually require prewetting with organic solvents before use. To do this, a column is picked up from a rack using the liquid distribution hand. The column is "parked" in a column-holding station, so that the hand can be removed and the column is left in a fixed position. An appropriate volume of an organic solvent is dispensed from the master lab station, through the nozzle, and into the column.
4. Add the sample and elute to waste: The liquid distribution hand detaches from the column and attaches a disposable pipet tip. It then transfers 0.500 ml of the prepared sample to the column and removes the pipet tip. The liquid distribution hand plugs into the column and elutes any undesired components in the sample to waste.
5. Elute the sample: The robot removes the column from the parking station and brings it to an autosampler vial or the inlet port of an LC injector station. The master lab station dispenses the appropriate volume of elution solvent, eluting the purified fraction of interest directly into the vial or loop. The next sample is prepared while the current one is being analyzed.

Automated ICP Analysis of Lubricating Oils

ICP is used for quantitative, simultaneous, multi-element analysis of samples such as catalysts, wastewaters, heavy oils, lubricating oils, and oil additives. Sample preparation for ICP typically requires weighing, liquid dispensing, mixing, heating, and pouring.

Donald Becker of Standard Oil of Indiana uses the Zymate Robotic System to prepare lubricating oil samples for wear metal analysis and to introduce the sample into the ICP.[5] If any sample falls outside the calibration range, the robotic system automatically dilutes and reruns the sample.

In this automated procedure the robot picks up an empty test tube, places it on a balance, and records the tare weight. Next, the robot picks up a test tube that contains the oil sample, heats it briefly using a heat gun, and pours the contents into the tared test tube. The actual sample weight is now calculated and the required amount of solvent is dispensed into the tube using the master lab station to achieve the desired concentration. The robot then places the tube into the vortex station where it mixes for 20 sec. The test tube is then removed and brought to the ICP "sipper" where the sample is drawn into the ICP. The "sipper" peristaltic pump is controlled by the power and event control station. The ICP is signaled to analyze the sample. If required, the robotic system automatically redilutes the sample when the calibration range is exceeded, if not it runs the next sample.

Nitrogen Analysis

Oil companies and catalyst manufacturers do a number of studies on the effect of processing oil and gasoline using new and different catalysts. One analysis requires monitoring the amount of nitrogen in each sample. The analytical measurement is performed by a chemiluminescence nitrogen analyzer. Typically, samples must be run the same day as received which makes this application a good candidate for automation.

In this procedure, the robot places a test tube on a balance and automatically records the tare weight. It then pours or pipets an oil or gasoline sample into the tared tube and calculates the actual sample weight. The master laboratory station adds the appropriate amount of toluene and based upon the sample weight the sample is mixed in the vortex station for a specified period of time. Using the syringe hand, the robot attaches a pipet tip, takes an aliquot of the sample, and puts it into a GC vial. The vial is brought to an air-actuated crimping station to be crimped and then loaded into an injector carousel.

Using a Robot in a Corrosive Environment

There are many applications that are desirable for a robot to perform, as the alternative is to have people in corrosive laboratory areas. One such application is metal digestion for atomic absorption spectroscopy (AA) or ICP, where it is necessary to use strong acids.

The mechanical and electronic components in the robot will need to be protected from the corrosive vapors. The changes and additions to be made to the robot are as follows:

1. The casing of the robot hand should be made from a corrosion-resistant polymer, such as polypropylene.
2. The finger racks on the general purpose hand, and the connector which attaches the robot wrist to the hand, should be made out of 316 gauge stainless steel rather than aluminum.
3. A gas purge inlet should be added to the robot hand. This gas purge vents into the wrist and rotects the electrical connectors between the hand and the wrist.
4. A gas purge should be added to the back of the robot arm and should also be designed to vent into the wrist.
5. A purge inlet should be added to the robot base to protect the mechanical and electronic assemblies located there.

It is highly recommended that there be adequate laboratory ventilation to minimize the buildup of corrosive vapors. Providing a flow of clean air past the robotic system before corrosive vapors are encountered will prolong the life of the system. It is also recommended that components such as the power and event control station and the master lab station and other system modules which may not be required in the actual vicinity of the robot, be placed away from the corrosive environment or, if necessary, be purged as well. In addition, the gas used to supply the robot purge inlets should be dry, so as to prevent condensation and subsequent corrosion of the purged components.

Routine Analysis of Formaldehyde in Aqueous Solution

The analysis of formaldehyde in aqueous solution is accomplished using the classic colorimetric method based on the formation of a deep purple complex formed between formaldehyde and para-rosanaline. The analysis involves a series of dilutions of the samples, adding para-rosanaline and sodium sulfite and waiting 1 hr for the color to develop. The intensity of the color, measured by a spectrophotometer, is directly proportional to the formaldehyde concentration. Formaldehyde concentrations in the sample are determined by comparison to a calibration curve.

The analytical laboratory at Standard Oil of Ohio has automated this procedure using a Zymate Laboratory Robotic System.[12]

The automated procedure starts with the operator manually placing the samples and empty test tubes into the appropriate racks and informing the system regarding the number of samples and dilution factors. The robot picks up a romote nozzle connected to the master lab station, and dispenses water into each of the test tubes. An aliquot of each sample is transferred to the first set of test tubes using a disposable pipet tip connected to the syringe hand. Mixing is accomplished by titration. A second dilution of each sample is made by transferring an aliquot from the first set of test tubes to the second set. The syringe hand then adds the para-rosanaline and the sodium sulfite to each sample. Reagent volumes equal one-tenth of the sample aliquot volume. The controller sets a timer for 1 hr after reagents are added to the first sample. After the incubation time elapses, the color intensity of each sample is measured by using a portable color probe attached to a **blank hand.** The analog output from the color probe is connected to the analog-to-digital converter built into the power and event control station. The sample concentration is automatically calculated from a previously prepared calibration curve. The calibration curve could also be generated at the time of analysis by including a set of standards with the samples. The results are then printed.

The automated method has typically reduced the error by one-half to one-third versus the manual procedure.

Automated Sample Introduction into a Superconducting NMR

It has been reported that Stan Gross of the research laboratories at Eastman Kodak Company has developed an automated sample introduction system for a Jeol FT-NMR Spectrometer using a Zymate Robotic system.[12] The advantages to a robotic approach to automating an NMR are:

- Less human labor required for repetitive operations
- Faster sample turnaround time
- Greater instrument utilization
- More flexibility in scheduling NMR experiments

The last three benefits are possible if the spectrometer operates overnight and on weekends.

The automated procedure begins with an operator placing the NMR tubes into spinners and then into the sample rack. The operator gives the computer the identification, solvent name, and number of experiments for each sample. The robot picks the spinner up from the rack and places it on the air cushion at the top of the superconducting magnet. Automated electronics lower the spinner into the magnet, obtain field/frequency lock, adjust magnet homogeneity, adjust NMR receiver gain, and collect the NMR data. The spinner is ejected from the magnet and the robot returns the spinner to the rack and then gets the next sample to run.

The Zymate robot was suitable for this application because it (1) can operate in a strong magnetic field; (2) does not degrade the magnetic homogeneity; and (3) does not produce (rf) interference in the NMR spectra.

Automated Tensile Testing of Rigid Polymers

The uniaxial tensile test is used by polymer manufacturers and fabricators to characterize the properties of developmental materials and to control quality during manufacture. This test determines the stress/strain relationship under a defined set of experimental conditions, from which tensile properties can be calculated.

Richard Scott and James Rieke of Dow Chemical's polymeric materials laboratory have reported using a Zymate Laboratory Robotic System to automatically introduce rigid polymer samples into an Instron Universal Tester.[13] In their procedure specimen samples are loaded into a verticle sample-dispensing rack and experimental conditions are entered into the laboratory computer. The robot then picks up a specimen from the dispenser and inserts it into the air-actuated Instron grips. The power and event control station signals the grips to close and begin the analysis. When the test is completed, the Instron sends a signal for the robot to remove the specimen and repeat the experiment with the sample.

Through the use of the robotic system, the throughput of the Instron Tester has essentially doubled.

Percent Solids, pH, and Brookfield Viscosity Measurement

A variety of simple but repetitive wet-chemistry tests are used throughout the chemcial industry for both product development and ongoing quality control. Typically these are labor intensive and may introduce operational delays while waiting for results.

Philip Mango of Air Products, uses a Zymate Laboratory Robotic System for automated measurement of pH, Brookfield viscosity, and percent solids of adhesive polymer emulsions.[14] A single robotic system performs all three tests in an integrated procedure.

The analysis begins with the robot placing a weighing pan on the balance and recording the tare weight. Using the syringe hand, the robot pipets an aliquot of the latex sample onto the weighing pan on the balance and records the actual sample weight. The robot closes the balance hood and initiates the infrared drying. While the sample is being dried, the robot brings a pH probe to the sample container to measure and record the pH of the sample. The pH meter is connected to the analog-to-digital converter built into the power and event control station. After the measurement is made, the pH probe is cleaned off in a washing station. The robot then brings the sample to the Brookfield viscometer and initiates the viscosity test. The output of the viscometer is also connected to analog-to-digital converter on the power and event control station. After the results are recorded, the robot brings another wash station to the viscometer spindle. The final weight of the dried sample is taken and the percent solids are calculated by the robot controller. Finally, all results are combined, printed, and transmitted to a host computer.

Automated Karl Fischer Titrations

Karl Fischer titration is one of the most widely performed titration procedures in the laboratory. Automated Karl Fischer titrators have been used throughout many industries for several years as the method of choice for determining both trace and major level moisture content. Through laboratory robotics, it is now possible to fully automate not only the titration step, but also the weighing, pretreatment, sample introduction, data compilation, and reporting.

The Karl Fischer titration method is used throughout the chemical industry for water analysis. Oils and oil additives are typically analyzed for trace moisture content while water-miscible organic solvents are typically analyzed for percent water using the same technique. Karl Fischer solvents and solvent combinations are diversified and versatile enough to accommodate a wide range of organic and inorganic samples including viscous oils, light oils, and organic solvents, as well as powdered and solid samples. The following procedure is typical for trace moisture analysis in viscous lubricating oil.

Solvent is added to the titration cell through a communication interface between the robotic controller and the titrator and an automated solvent dispenser. Acetic acid is added to the titration cell through the master lab station. A signal from the controller to the titrator begins the titration with the Karl Fischer reagent. Next, the sample is weighed and added to the titration cell using the general purpose hand, a balance interface, and a **viscous liquid handling kit.** The viscous liquid handling kit is a disposable device that allows the transfer of viscous samples to the Karl Fischer cell with minimal exposure to air. The sample is then automatically titrated with the Karl Fischer reagent and the titrator. The results are calculated, and printed or sent on to a host computer.

Completely automated Karl Fischer titrations, including sample weighing and reagent additions, can be run unattended using a laboratory robotic system integrated with a Karl Fischer titrator. Laboratory productivity is increased by freeing a technician from continually weighing samples and introducing them manually into the automatic titrator. In addition, the productivity and utilization of the titrator can be doubled or tripled by operating unattended for up to 24 hrs per day.

Automated Spectroscopy; Determination of Polyacrylamide (PAM) in Water

Spectrophotometric analyses are widely used throughout the chemical industry. Effective sample preparation techniques can remove interferences and enhance detection-making spectroscopic procedures by making them more precise and reliable.

Robert Kirsch and Virginia Lang of American Cyanamid have used a Zymate Robotic System to automate sample preparation for a spectrophotometric procedure.[15] The procedure determines PAM in water at the 1 to 15 ppm range. Detection is enhanced by reacting the PAM amide groups with bromine. The bromine oxidizes iodine ions to form iodide which is measured as a starch-tri-iodine complex.

In this procedure the robot attaches a 5 ml pipet tip, and, using the master lab station, pipets 5 ml of sample into a vial on a shaker. Using the syringe hand, 1 ml of acetate buffer followed by 0.2 ml of bromine water is added. The shaker is automatically turned on and shakes for 15 min (the shaker is modified to stop and start in the same place each time). The shaker stops at the appropriate time, and 3 ml of starch-cadmium iodide solution is added, and the mixture is shaken for another 15 min. The robot attaches the general purpose hand, picks up the first sample vial, and brings it to a peristaltic sipper pump inlet on the spectrophotometer.

The sample is drawn into the flow cell by having the power and event control station turn on the sipper pump on the spectrophotometer. The data from the spectrophotometer is printed or sent on to a host computer. The next sample flushes out the previous sample from the flow cell.

There are four general approaches to automatic sample introduction for spectroscopy. In each case the power and event control station signals the spectrophotometer to make the measurement and then monitors the analog absorbance response from the spectrophotometer.

1. Cuvette insertion. Direct insertion and removal of the cuvette by the robot.
2. Syringe dispensing into an open cuvette. Direct sample introduction into a preplaced cuvette using the syringe hand.
3. Interface to automated spectrophotometer. The sample vial is presented by the robot to the flow or sipper inlet of the spectrophotometer.
4. Automated spectrophotometer interface. The master lab station automatically introduces and removes samples. For example, a sample vial is presented to the inlet of a syringe in the master lab station. A sample aliquot is drawn into the syringe with the syringe valve in an initial position. The valve is automatically switched to another position and the sample is dispensed into the cell. When the measurement is complete, the sample is drawn back into the syringe, the valve is switched back to the initial position, and the syringe returns the sample to the original vial.

Sample Preparation for X-ray Fluorescence

Many industrial laboratories use x-ray fluorescence (XRF) for the analysis of major and minor constituents of oils, ores, slags, cements, steel, aluminum, and other alloys.

The general sample preparation procedure for ores and slags begins when the robot attaches the powder dispensing hand, places a container on the balance, and tares it through the balance interface. The robot then picks up a sample container and weighs out the desired amount of the powdered sample. Based on this weight, the robot controller calculates the amount of flux (usually powdered lithium metaborate) to be added. This can be done with the powder dispensing hand or a reagent powder dispenser such as the Heirath and Andrews ISO-G Direct Weight Filler, which is controlled by the power and event control station. The sample container is transferred to a vortexor to mix the sample and the flux. The mixture is then placed into a crucible, and the crucible is manually placed into a fusion device.

Tables 4-1 and 4-2 illustrate some uses of laboratory robotics. Table 4-1 enumerates various laboratory operations that have been automated with robotics. Table 4-2 provides a categorization of applications in several generalized areas.

TABLE 4-1. Applications That Have Been Automated Using Robotics

Dissolution testing (both endpoint and multipoint)
Total nonvolatile extractable analysis
Limulus amebocyte lysate (LAL) procedure for the detection of bacterial endotoxin
Extraction prodedure for tricyclic antidepressants in clinical monitoring
Content uniformity testing
DNA-ninhydrin assay
Tablet dissolution assay
Spiral streaking of petri dishes
Robotic optimization of organic reactions
Radioligand binding assays
Robotics in sterile tissue culture techniques
Automated feed analysis
Physical testing of paper and other sheeted material
Estimation of Hepatitis B surface antigen concentration
Determination of drugs in physiological samples
Pharmaceutical capsule production
Introduction of samples into emission spectrophotometer
Sample preparation of corn tissue for nitrate analysis
Carbon Residue Analysis
Polymer Moisture Analysis
Blood Typing using **bar coded labels**
Tensile testing of nonwoven fiber samples
Elemental analysis of semiconductors
Radioimmunoassay for T4
Allergen formulation testing
Quality control of immunodiagnostic kits
Environmental analysis of priority pollutants by GC
Automated titrations of isocyanates (ELISA) assays
General sample preparation for drug metabolism assays
Pesticide residue sample preparation for GC
Sample preparation of fermentation broths for HPLC
Automated methods development
Microplate testing and reading
Pulsed NMR
Sample changer for superconducting NMR
Cleaning of NMR tubes
Paint formulation
Specific ion electrode measurements
X-ray fluorescence (powders and oils)
Preservative testing
Metal digestion preparation for ICP
Flavor analysis by GC
Feed analysis for toxicology
Sample introduction of polymer samples into an Instron
Soil analysis
Sample preparation for light-scattering spectroscopy
Photographic emulsion research
pH measurement and endpoint titration
Viscosity measurement of polymers
Total solid determination of polymers
Suspended solid determination of environmental samples

TABLE 4-1. (continued)

- Oil additive formulation
- Oil blending
- Sample preparation for GPC
- Toxicology sample preparation for HPLC
- Genetic engineering
- Vitamin A analysis preparation for HPLC
- DNA purification
- General sample preparation for food analysis by HPLC
- Sample preparation of caffeine analysis
- Clean room solvent wash operation
- Multi-vitamin assay for HPLC
- Enzyme kinetics
- Pesticide formulation analysis
- Thermal analysis (DSC)
- Autosampler for neutron activation analyzer
- Sample preparation for (IR) analysis
- Spectrophotometric analysis of polymer dyes
- Diethyleneglycol analysis of polymer samples for GC
- BOD determinations
- Calorimetric procedure and weighing
- Nitrogen analysis of chemiluminescence
- New product formulation
- Paraben sample preparation
- Tissue residue analysis
- Radioiodination analysis
- Suspended solids in waste water
- Dust filter weighing
- Serial dilutions
- (LASA-P) cancer monitoring test
- Physical testing of paper products
- Ion chromatography preparation and introduction
- Head space analysis
- Extraction and sampling for a technicon autoanalyzer
- Radioactive waste manipulation
- Acid digestion procedure
- Wear metals in oil for ICP
- Contact lens solution testing
- Lowry protein assay
- Bottle-washing for AA
- Colorimetric endpoint titration
- (Bio-Rad) protein assay
- Liquid/liquid extractions
- Liquid/solid extractions
- Karl Fischer titrations
- Analytical analysis of parasites
- Cholesterol analysis
- Gamma counting
- Catalyst decomposition
- Fluoride analysis in toothpaste
- Cellulose hydrolysis
- Polarography
- Thin layer chromatography
- Hazardous work environments
- Microplate manipulation
- Hybridoma research
- Stability testing
- Sugar analysis by HPLC
- Environmental evaluation of water conditions using metal coupons
- Particle-size analysis
- Microbiological inoculation and mixing of cosmetic preservation testing
- Preparation of herbicide samples for HPLC
- Industrial hygiene preparation using extraction tubes
- Sample preparation for plutonium and americium radiochemical analysis
- Hydrogen and nitrogen analysis
- Weighing and calorimetry for production of plutonium-238 oxide-fueled milliwatt generators
- Preparation of fused beads for XRF

TABLE 4-2. Selected Applications of Laboratory Robotics[a]

Pharmaceutical analysis (31, 34, 35, 36, 39, 41, 51, 60, 61, 70)
Pharmaceutical dissolution testing (30, 40, 71)
General pharmaceutical (32, 47, 60, 62)
HPLC and GC (1, 4, 5, 23, 33, 50, 52, 69, 75)
ICP and AA (29, 43, 76)
Materials science (10, 11, 24, 42, 53, 54, 66)
Drug and metabolite analysis (14, 15, 19, 45, 46, 49, 51, 59)
Sample preparation (6, 7, 14, 16, 19, 25, 27, 28, 34, 36, 37, 39, 50, 76)
X-ray fluorescence (8, 9, 73)
Organic synthesis (12, 44, 74)
Oligonucleotide purification (13)
Radioiodinations (17)
Environmental (5, 18, 56)
Food and nutrition (20, 21, 22, 50)
Microbiology (38, 48, 58)
General laboratory (2, 28, 41, 55, 57, 63, 64, 67, 72)
Titration (26, 65)
UV/Vis Spectroscopy (68)
NMR (3)
Agricultural (4)

[a]Number refers to Applications Bibliography entries.

REFERENCES

1. Van Antwerp, J.; Venteicher, R.F. Improving the flexibility of an analytical robotic system by use of programmable column switching, solvent selections, and robotic computer control programmable HPLC equipment. In "Advances in Laboratory Automation Robotics 1985"; Hawk, G.L.; Strimaitis, J.R., Eds.; Zymark, Hopkinton, Massachusetts, 1985; p 75.
2. Addison, J.H.; Dyches, G.M. A shared robotic system: Automated pipette calibration and pipette tip filter assembly. In "Advances in Laboratory Automation Robotics 1985"; Hawk, G.L.; Strimaitis, J.R., Eds.; Zymark, Hopkinton, Massachusetts, 1985; p 87.
3. Ragouzeos, A.; Crouch, R.; Miller, J.L. Interface of Zymark robot with Varian FT80 NMR spectrometer. In "Advances in Laboratory Automation Robotics 1985"; Hawk, G.L.; Strimaitis, J.R., Eds.; Zymark, Hopkinton, Massachusetts, 1985; p 105.
4. Goldberg, S.S.; Preparation of herbicide samples for HPLC analysis by robotics. In "Advances in Laboratory Automation Robotics 1985"; Hawk, G.L.; Strimaitis, J.R., Eds.; Zymark, Hopkinton, Massachusetts, 1985; p 105.
5. Haile, D.M.; Brown, R.D. Robotic preparation of water samples for trace herbicide analysis. In "Advances in Laboratory Automation Robotics 1985"; Hawk, G.L.; Strimaitis, J.R., Eds.; Zymark, Hopkinton, Massachusetts, 1985; p 123.
6. Knobeloch, D.W.; Austin, L.R.; Latimer, T.W.; Schneider, D.N. Automation of the calorimetry step for production of plutonium-238-oxide-fueled milliwatt genera-

tors. In "Advances in Laboratory Automation Robotics 1985"; Hawk, G.L.; Strimaitis, J.R., Eds.; Zymark, Hopkinton, Massachusetts, 1985; p 313.
7. Hilliard, L.J.; Alexakos, L.G.; Kobrin, R.J.; Granchi, M.P.; Grey, P. Automated hydrogen and nitrogen analyses using a Zymark robot. In "Advances in Laboratory Automation Robotics 1985"; Hawk, G.L.; Strimaitis, J.R., Eds.; Zymark, Hopkinton, Massachusetts, 1985; p 325.
8. Cross, J.B.; Wilson, L.V.; Marak, E.J.; Jones, R.D. Robotic automation for X-ray fluorescence analysis of sulfur in oils. In "Advances in Laboratory Automation Robotics 1985"; Hawk, G.L.; Strimaitis, J.R., Eds.; Zymark, Hopkinton, Massachusetts, 1985; p 347.
9. Petin, J.; Wagner, A. Automatic preparation of fused beads for X-ray fluorescence analysis by the combination of a Perl'X2 bead machine with a Zymark Robotic System. In "Advances in Laboratory Automation Robotics 1985"; Hawk, G.L.; Strimaitis, J.R., Eds.; Zymark, Hopkinton, Massachusetts, 1985; p 367.
10. Gateff, P.A.; Abbott, J.C. Laboratory robotics: Applications in the materials science laboratory. In "Advances in Laboratory Automation Robotics 1985"; Hawk, G.L.; Strimaitis, J.R., Eds.; Zymark, Hopkinton, Massachusetts, 1985; p 379.
11. Koeninger, E.C.; Grano, J.; Heaps, J.F. Formulation and testing for a coating application for research and development. In "Advances in Laboratory Automation Robotics 1985"; Hawk, G.L.; Strimaitis, J.R., Eds.; Zymark, Hopkinton, Massachusetts, 1985; p 407.
12. Kramer, G.W.; Fuchs, P.L. Robotic automation in organic synthesis. In "Advances in Laboratory Automation Robotics 1985"; Hawk, G.L.; Strimaitis, J.R., Eds.; Zymark, Hopkinton, Massachusetts, 1985; p 417.
13. Jones, S.S.; Brown, J.E.; Vanstone, D.A.; Stone, D.; Brown, E.L. Synthetic DNA: Application of robotics to the purification of oligonucleotides. In "Advances in Laboratory Automation Robotics 1985"; Hawk, G.L.; Strimaitis, J.R., Eds.; Zymark, Hopkinton, Massachusetts, 1985; p 431.
14. Castellani, W.J.; Pippenger, C.E.; Galen, R.S. Robotic sample preparation for automated batch-oriented analysis in the clinical chemistry laboratory. In "Advances in Laboratory Automation Robotics 1985"; Hawk, G.L.; Strimaitis, J.R., Eds.; Zymark, Hopkinton, Massachusetts, 1985; p 449.
15. Kramer, S.F.; Levitt, M.J.; Passarello, M.M. Comparison of automated and manual extraction of drugs from biological fluids at trace levels. In "Advances in Laboratory Automation Robotics 1985"; Hawk, G.L.; Strimaitis, J.R., Eds.; Zymark, Hopkinton, Massachusetts, 1985; p 465.
16. Brennan, J.E.; Severns, M.L.; Kline L.M. Centralized sample preparation using a laboratory robot. In "Advances in Laboratory Automation Robotics 1985"; Hawk, G.L.; Strimaitis, J.R., Eds.; Zymark, Hopkinton, Massachusetts, 1985; p 481.
17. Hurni, W.M.; Wasmuth, E.D.; Miller, W.J.; McAleer, W.J. A robot for performing radioiodinations. In "Advances in Laboratory Automation Robotics 1985"; Hawk, G.L.; Strimaitis, J.R., Eds.; Zymark, Hopkinton, Massachusetts, 1985; p 497.
18. Kropscott, B.E.; Coyne, L.E.; Campbell, R.A.; Sowle, W.F. Robotic applications within Dow's health and environmental sciences laboratory. In "Advances in Laboratory Automation Robotics 1985"; Hawk, G.L.; Strimaitis, J.R., Eds.; Zymark, Hopkinton, Massachusetts, 1985; p 131.

19. Rollheiser, J.J.; Schmidt, W.A.; Stelting, K.M. Laboratory robotics applied to chemistry for toxicology. In "Advances in Laboratory Automation Robotics 1985"; Hawk, G.L.; Strimaitis, J.R., Eds.; Zymark, Hopkinton, Massachusetts, 1985; p 149.
20. Dulitzky, M. Development of a robotic caffeine analysis. In "Advances in Laboratory Automation Robotics 1985"; Hawk, G.L.; Strimaitis, J.R., Eds.; Zymark, Hopkinton, Massachusetts, 1985; p 163.
21. Lento, H.G.; Grady, M.D.; Hastings, H.J. The role of robotics in the automated determination of the nutritional composition of foods—A progress report. In "Advances in Laboratory Automation Robotics 1985"; Hawk, G.L.; Strimaitis, J.R., Eds.; Zymark, Hopkinton, Massachusetts, 1985; p 179.
22. Higgs, D.J.; Vanderslice, J.T.; Huang, M.A. Automated robotic extraction and subsequent analysis of vitamins in food samples. In "Advances in Laboratory Automation Robotics 1985"; Hawk, G.L.; Strimaitis, J.R., Eds.; Zymark, Hopkinton, Massachusetts, 1985; p 195.
23. Markelov, M.; Antloga, M.; Schmidt, S.A. Automation of multiple procedures in an industrial laboratory (trace organics in water and soil, residual monomers, anionic surfactants, preparation of standards for GC, LC & IE). In "Advances in Laboratory Automation Robotics 1985"; Hawk, G.L.; Strimaitis, J.R., Eds.; Zymark, Hopkinton, Massachusetts, 1985; p 75.
24. Prozonic, F.M. Dual function robotics system: Autosampler for thermal analysis and applications in corrosion studies. In "Advances in Laboratory Automation Robotics 1985"; Hawk, G.L.; Strimaitis, J.R., Eds.; Zymark, Hopkinton, Massachusetts, 1985; p 231.
24. Klinger, K.A. Automated sample preparation procedures for liquid and gas chromatographic analysis of polymeric materials. In "Advances in Laboratory Automation Robotics 1985"; Hawk, G.L.; Strimaitis, J.R., Eds.; Zymark, Hopkinton, Massachusetts, 1985; p 247.
26. Simonson, L.A. Laboratory robotics: An application in automated titrations. In "Advances in Laboratory Automation Robotics 1985"; Hawk, G.L.; Strimaitis, J.R., Eds.; Zymark, Hopkinton, Massachusetts, 1985; p 269.
27. Beugelsdijk, T.J.; Knobeloch, D.W.; Thurston, A.A.; Stalnaker, N.D.; Austin, L.R. Robot-assisted sample preparation for plutonium and americium radiochemical analysis. In "Advances in Laboratory Automation Robotics 1985"; Hawk, G.L.; Strimaitis, J.R., Eds.; Zymark, Hopkinton, Massachusetts, 1985; p 283.
28. Jones, R.D.; Cross, J.B. Automating sample preparation (and disposal) with a robotic workcell. In "Advances in Laboratory Automation Robotics 1985"; Hawk, G.L.; Strimaitis, J.R., Eds.; Zymark, Hopkinton, Massachusetts, 1985; p 293.
29. Lester, L.; Lincoln, T.; Donoian, H. Development of an automated urine-analysis scheme for determination of ppb levels of As and Se via hydride/atomic absorption. In "Advances in Laboratory Automation Robotics 1985"; Hawk, G.L.; Strimaitis, J.R., Eds.; Zymark, Hopkinton, Massachusetts, 1985; p 509.
30. Compton, B.J.; Zazulak, W.; Hinsvark, O. Robotic sample retrieval from pharmaceutical dissolution testers. In "Advances in Laboratory Automation Robotics 1985"; Hawk, G.L.; Strimaitis, J.R., Eds.; Zymark, Hopkinton, Massachusetts, 1985; p 531.
31. Walsh, P.; Abdou, H.; Barnes, R.; Cohen, B. Laboratory robotics for tablet content uniformity testing. In "Advances in Laboratory Automation Robotics 1985";

Hawk, G.L.; Strimaitis, J.R., Eds.; Zymark, Hopkinton, Massachusetts, 1985; p 547.
32. Hall, M.A.; Kiral, R.M.; Dziabo, A.J.; Zymate laboratory automation system in contact lens product research and development in laboratory. In "Advances in Laboratory Automation Robotics 1985"; Hawk, G.L.; Strimaitis, J.R., Eds.; Zymark, Hopkinton, Massachusetts, 1985; p 563.
33. Halloran, K.J.; Franze, H.M. Interaction between a robotic system and liquid chromatograph—HPLC control, communication and response. In "Advances in Laboratory Automation Robotics 1985"; Hawk, G.L.; Strimaitis, J.R., Eds.; Zymark, Hopkinton, Massachusetts, 1985; p 575.
34. Hatfield, C.; Halloran, E.; Habarta, J.; Romano, S.; Mason, W. Multi-product sample preparation in the pharmaceutical quality assurance laboratory. In "Advances in Laboratory Automation Robotics 1985"; Hawk, G.L.; Strimaitis, J.R., Eds.; Zymark, Hopkinton, Massachusetts, 1985; p 599.
35. Hatton, B.; Abley, P.; Lux, T.J. The extension of pharmaceutical analysis automation using robotics. In "Advances in Laboratory Automation Robotics 1985"; Hawk, G.L.; Strimaitis, J.R., Eds.; Zymark, Hopkinton, Massachusetts, 1985; p 621.
36. Johnson, J.H.; Srinivas, R.; Kinzelman, T.J. Automated sample preparation of pharmaceutical parenterals for analysis using robotics. In "Advances in Laboratory Automation Robotics 1985"; Hawk, G.L.; Strimaitis, J.R., Eds.; Zymark, Hopkinton, Massachusetts, 1985; p 637.
37. Kilbourn, M.R.; Brodack, J.W.; Welch, M.J.; Katzenellenbogen, J.A. Application of robotics for the routine production of fluorine-18-labeled radiopharmaceuticals. In "Advances in Laboratory Automation Robotics 1985"; Hawk, G.L.; Strimaitis, J.R., Eds.; Zymark, Hopkinton, Massachusetts, 1985; p 663.
38. Smith, J.L. Use of the Zymate robot for microbiological inoculation and mixing of cosmetic preservation testing samples. In "Advances in Laboratory Automation Robotics 1985"; Hawk, G.L.; Strimaitis, J.R., Eds.; Zymark, Hopkinton, Massachusetts, 1985; p 677.
39. Inman, G.W.; Elks, D.D. General purpose robotic preparation of composite tablet samples for HPLC analysis. In "Advances in Laboratory Automation Robotics 1985"; Hawk, G.L.; Strimaitis, J.R., Eds.; Zymark, Hopkinton, Massachusetts, 1985; p 689.
40. Kostek, L.J.; Brown, B.A.; Curley, J.E. Fully automated dissolution testing through robotics. In "Advances in Laboratory Automation Robotics 1985"; Hawk, G.L.; Strimaitis, J.R., Eds.; Zymark, Hopkinton, Massachusetts, 1985; p 701.
41. Greenberg. A.; Young, R. Totally automated robotic procedure for assaying composite samples which normally require large volume dilutions. In "Advances in Laboratory Automation Robotics 1985"; Hawk, G.L.; Strimaitis, J.R., Eds.; Zymark, Hopkinton, Massachusetts, 1985; p 721.
42. Mango, P.A. Robotics in polymer testing and characterization. "Advances in Laboratory Automation Robotics 1984"; Hawk, G.L.; Strimaitis, J.R., Eds.; Zymark, Hopkinton, Massachusetts, 1984; p 17.
43. Becker, D.A. Robot control of ICP analyses of lubricating oils and lubricating oil additives. In "Advances in Laboratory Automation Robotics 1984"; Hawk, G.L.; Strimaitis, J.R., Eds.;Zymark, Hopkinton, Massachusetts, 1984; p 35.
44. Kramer, G.W.; Frisbee, A.R.; Fuchs, P.L. Robotic optimization of organic reactions—Phase I. In "Advances in Laboratory Automation Robotics 1984"; Hawk, G.L.; Strimaitis, J.R., Eds.; Zymark, Hopkinton, Massachusetts, 1984; p 47.

45. Schoenhard, G.; Schmidt, R.; Kosobud, L.; Smykowski, K. Robotics assays for drugs in animal and human plasma. In "Advances in Laboratory Automation Robotics 1984"; Hawk, G.L.; Strimaitis, J.R., Eds.; Zymark, Hopkinton, Massachusetts, 1984; p 61.
46. Myers, D.J.; Szuminsky, N.; Levitt, M.J. Automating drug metabolism studies through laboratory robotics. In "Advances in Laboratory Automation Robotics 1984"; Hawk, G.L.; Strimaitis, J.R., Eds.; Zymark, Hopkinton, Massachusetts, 1984; p 71.
47. Taylor, J.E. Pharmaceutical applications of laboratory robotics: Radioligand binding assays. In "Advances in Laboratory Automation Robotics 1984"; Hawk, G.L.; Strimaitis, J.R., Eds.; Zymark, Hopkinton, Massachusetts, 1984; p 83.
48. Hupe, D.; Peters, K. The use of robotics in sterile culture techniques. In "Advances in Laboratory Automation Robotics 1984"; Hawk, G.L.; Strimaitis, J.R., Eds.; Zymark, Hopkinton, Massachusetts, 1984; p 91.
49. Kropscott, B.E.; Dittenhafer, M.L. Automated feed analysis—An application for toxicology. In "Advances in Laboratory Automation Robotics 1984"; Hawk, G.L.; Strimaitis, J.R., Eds.; Zymark, Hopkinton, Massachusetts, 1984; p 105.
50. Hurst, W.J. Laboratory robotics applied to food analysis. In "Advances in Laboratory Automation Robotics 1984"; Hawk, G.L.; Strimaitis, J.R., Eds.; Zymark, Hopkinton, Massachusetts, 1984; p 117.
51. Pheil, M.J.; Walden, G.L.; Weismuller, J.C. Automation of U.S.P. glycerine analyses by the use of a robotics system. In "Advances in Laboratory Automation Robotics 1984"; Hawk, G.L.; Strimaitis, J.R., Eds.; 1984; Zymark, Hopkinton, Massachusetts, p 125.
52. Antloga, M.; Markelov, M.; Pagliaro, L. Novel approaches to solving a variety of problems in an industrial analytical chemistry laboratory. In "Advances in Laboratory Automation Robotics 1984"; Hawk, G.L.; Strimaitis, J.R., Eds.; Zymark, Hopkinton, Massachusetts, 1984; p 137.
53. Scott, R.L.; Rieke, J.K. Development of a fully automated tensile test system. In "Advances in Laboratory Automation Robotics 1984"; Hawk, G.L.; Strimaitis, J.R., Eds.; Zymark, Hopkinton, Massachusetts, 1984; p 151.
54. McLaughlin C.; Abbott, J.C.; Jenkins, L.A. Laboratory robotics in the physical testing of paper and other sheeted materials. In "Advances in Laboratory Automation Robotics 1984"; Hawk, G.L.; Strimaitis, J.R., Eds.; Zymark, Hopkinton, Massachusetts, 1984; p 165.
55. Cross, J.B.; Marak, E.J. Automated dispensing and weighing of solid powders using robotics. In "Advances in Laboratory Automation Robotics 1984"; Hawk, G.L.; Strimaitis, J.R., Eds.; Zymark, Hopkinton, Massachusetts, 1984; p 181.
56. Kirsch, R.B.; Lang, V.B. The determination of trace levels of polyacrylamide in water using robotics. In "Advances in Laboratory Automation Robotics 1984"; Hawk, G.L.; Strimaitis, J.R., Eds.; Zymark, Hopkinton, Massachusetts, 1984; p 193.
57. Wasmuth, E.H.; Aiello, P.J.; Miller, W.J.; McAleer, W.J. Estimation of hepatitis B surface antigen concentration using a Zymate laboratory robotic system. In "Advances in Laboratory Automation Robotics 1984"; Hawk, G.L.; Strimaitis, J.R., Eds.; Zymark, Hopkinton, Massachusetts, 1984; p 209.
58. Martin, P.A.; Tsuji, K. Automation of bacterial endotoxin testing with a Zymate robotic system. In "Advances in Laboratory Automation Robotics 1984"; Hawk, G.L.; Strimaitis, J.R., Eds.; Zymark, Hopkinton, Massachusetts, 1984; p 219.

59. Lewis, E.C.; Santarelli, D.R.; Malbica, J.O. Laboratory robotics for automated determination of drugs in physiological samples. In "Advances in Laboratory Automation Robotics 1984"; Hawk, G.L.; Strimaitis, J.R., Eds.; Zymark, Hopkinton, Massachusetts, 1984; p 237.
60. Siebert, E. An application of robotics to pharmaceutical tablet sample preparation. In "Advances in Laboratory Automation Robotics 1984"; Hawk, G.L.; Strimaitis, J.R., Eds.; Zymark, Hopkinton, Massachusetts, 1984; p 257.
61. Venteicher, R.F.; Van Antwerp, J. Implementation and validation of a robotic system for quality control testing. In "Advances in Laboratory Automation Robotics 1984"; Hawk, G.L.; Strimaitis, J.R., Eds.; Zymark, Hopkinton, Massachusetts, 1984; p 175.
62. Compton, B.; Froome, M.; Hinsvark, O. A robotically assisted system for pharmaceutical capsule production. In "Advances in Laboratory Automation Robotics 1984"; Hawk, G.L.; Strimaitis, J.R., Eds.; Zymark, Hopkinton, Massachusetts, 1984; p 287.
63. Auses, J.P.; McCutcheon, K.R.; Smith, J.W. Automated analytical chemistry at the Alcoa technical center. In "Advances in Laboratory Automation Robotics 1984"; Hawk, G.L.; Strimaitis, J.R., Eds.; Zymark, Hopkinton, Massachusetts, 1984; p 311.
64. Severns, M.L.; Brennan, J.E. The use of robotics in bar code label testing and evaluation. In "Advances in Laboratory Automation Robotics 1984"; Hawk, G.L.; Strimaitis, J.R., Eds.; Zymark, Hopkinton, Massachusetts, 1984; p 323.
65. Brown, R.K.; Volk, P.R.; Scharicz, K.E. Fully automated titrations using laboratory robotics, Pittsburgh Conference, 1985; 577.
66. Abbott, J.C.; McLaughlin, C.A.; Jenkins, L.A. Automated robot-controlled analysis of sheeted materials for physical characterization, Pittsburgh Conference on Analytical Chemistry and Applied Spectroscopy, Atlantic City, New Jersey, 1985; 579.
67. Tryon, P.J. Robotic sample handling instrumentation, Pittsburgh Conference on Analytical Chemistry and Applied Spectroscopy, Atlantic City, New Jersey, 1985; 582.
68. Williams, J.F.; McGrattan, P.A. Robotic automation of UV/VIS spectrophotometric bioassays, *Instrumentation Research* **1986**.
69. Vivilecchia, R.V. Integration of laboratory robotics and HPLC, *LC Magazine*, **1986**, 4, 94.
70. Strimaitis, J.R., Pharmaceutical quality control using laboratory robotics, *American Laboratory*, **1986**, 28.
71. Dolan, T.; Saboe, T.; Sattler, L. Automation of Dissolution Testing by Robotics and Computerized Data Reduction, presented at APHA, 131st Annual Meeting, 1984.
72. Macero, D.J. Robotic tools for flexible laboratory automation, Eastern Analytical Symposium, New York City, 1985; 76.
73. Seiler, B.D. The application of robotics for X-ray fluorescence sample preparation, EAS, 1985; 77.
74. Lochmuller, C.H.; Lung, K.R. The laboratory robot as an explorer of chemical reactions: factor analysis, optimization and response surfaces, EAS, 1985, 785.
75. Martin, M.W.; Rao, J.M. Automated determination of extractable nitrates in corn plant tissue by laboratory robotics/ion chromatography, EAS, 1985, 116.
76. Sharicz, K.E.; Brown, R.E.; Volk, P.A. Automated sample preparation for atomic spectroscopy, Pittsburgh Conference 1985, 574.

5
Getting Started in Laboratory Robotics

EXPECTATIONS

What are your expectations in regard to acquiring a laboratory robotic system? Since this is a relatively new technology, it can be initially difficult to measure whether your expectations are realistic. The following questions may help you focus on realistic expectations.

What to you really expect the robot system to do? On the surface this sounds like a ridiculous question. You expect it to run your samples, but how soon do you expect it to be up and running? You should expect to be able to run real samples about 6 weeks after receiving it. This will only happen through careful planning. Some people have said that what they want is a "turnkey" or "custom-engineered system," so all they have to do is plug the robotic system in, push a button, and it will run samples. Even with a turnkey system approach it is doubtful that you will be able to run samples and trust the results sooner than if you had worked on the system yourself, as you will have to rely upon an outside group to provide exactly what you want and to make any changes as they come up. This can be very time-consuming, frustrating, and expensive. It is best for you to work with and program your own robotic system with the support of the vendor or an internal support group. You will become self-sufficient, able to make changes when needed, and will have realistic expectations. You understand your particular application best and the vendor of a robotic system knows his system; by establishing a partnership you can both work toward a common goal.

What does your boss expect? Your boss may need to be informed. Nothing is worse than to have the boss keep asking "Well, is that robot up and running yet?" He should know beforehand that it will take several weeks before the system is fully operational. He should also provide the time for you or someone in your organization to work on it.

What do your colleagues expect? Your colleagues might expect the robot to move much faster than what they initially see. A very common comment is, "I can beat the robot, I can do it faster." They are right for the first three samples, and then the robot begins to win. Contests to see who is faster, man or robot, should not be encouraged. People should know weeks in advance, before the robot shows up, where it is going and what it will be doing. No one likes surprises. They should know what they will be working on and that the robot is a very positive thing and not a threat. The robot will provide them with more, not fewer, opportunities.

How much time can you initially devote to the robot? Working on the robot part time or when you get a free minute will not work. It is important that you make the commitment to get the time to work on the robot. If you don't, the project will take too much time to get up and running. It was important to buy the robot and it is important to make the time to work on it.

Where will you put it in the laboratory? Most laboratories are designed for people, not robots. Don't wait until the robot arrives to decide where it should go.

Have you assigned someone to the robotic project? Initially someone should be in charge of the robot and responsible for the completion of the project. That individual must be given the time to work on it. It is also important that they really want to work on the robot.

Is the actual user of the robot going to attend a school run by the manufacturer? The actual user should go to a school provided by the vendor before the robotic system arrives. Don't send the supervisor to school and expect him to teach the end-user how to program the system.

How many samples do you expect to run in a day? The number must be able to match the capabilities of the system. It may be able to run only so many samples before human intervention is required. The number of samples to be run in a day should be part of the initial planning.

When do you expect to be able to run real samples? Several factors have to be considered before a specific time for programming the robotic system can be estimated. These include: (1) the complexity of the task to be accomplished; (2) the capabilities of the robot and its peripherals; and (3) the capabilities of the person doing the programming. It will take much longer to teach a robot to perform a long, complicated application as opposed to a simple application. Programming time can range from a few days to several months; but on the average, most applications are complete in 6 weeks. Once the procedure is programmed, it can be stored indefinitely. Some people will start off with simple programs and continue to add on to their procedure. Initially the robot may only be programmed to weigh out the sample and dilute it. Once that has been accomplished, shaking, extracting, filtering, and injecting into the HPLC may be added to the program. If you were to hire a new chemist, how long would it be until you would rely on his answers on real samples?

How fast do you expect the robot to run the samples? It is unrealistic to expect that the robot will run samples much faster than people do. In some instances the robot will be slower. As a general rule, expect that the robot will process samples at about the same rate as people do.

Do you expect the robot to run the samples in exactly the same way as people do manually? If you want the robot to run samples exactly the way people do, you will probably be disappointed and will not be able to fully utilize the robot's capability. Experiments, glassware, hardware, and procedures were designed for people, not robots. Without changing the chemistry, most procedures can be changed to optimize the performance of the robot and the quantity of sample throughput. People work in a batch mode; robots work best in a serial mode. Separatory funnels are good for people but not for a robot, a robot could perform the liquid/liquid extraction better in a centrifuge tube. Be flexible and change the experimental technique, not the chemistry.

Is your application well defined? The application must be well defined. It is difficult to hit a moving target. If there are problems with the chemistry, the robot will not make them go away, it will probably amplify them.

Do you have the proper power requirements? Most laboratory robots don't need anything special as far as power is concerned, but the power line should be clean and reasonably free of spikes on the line. Power fluctuations could cause difficulties for the robot.

Will data be transferred to another computer? If data is going to be transferred to a host computer, make sure that the robotic system is compatible with it and that there is someone in your organization who can do the program and the interfacing job.

Have the disposable items been checked out to ensure they are "robot friendly," reproducibly made, and readily available? Disposable items have been designed for people, not robots. This has been true in any type of transfer from manual labor to machine. For example, if you wanted to keep two pieces of paper together, you would use a paper clip to do it manually, to automate that process no one tried to make an automated paper clip attacher, they invented the stapler. The staple is easy for the machine to deal with but extremely difficult for people to use. Disposable items in the laboratory will change in time for robotics, in the meantime it is important to check out the availability, reproducibility, and robot friendliness of these lab items.

Are the other pieces of equipment that the robotic system will interface with friendly and compatible? There are a number of pieces of equipment in the laboratory that in their present form cannot be used with a robotic system. Just because equipment has a binary coded decimal (BCD) outlet or an RS-232 port, doesn't make it robot compatible. Even if it is electronically compatible, it also must be mechanically compatible. For example, a centrifuge can easily be turned on and off and the speed varied, but opening the door and stopping the rotor in a predetermined location each time is a difficult job. Fortunately, manufacturers are making "robot friendly" devices for the laboratory, such as centrifuges, balances, shakers, etc. Have the robotic vendor check out the pieces of equipment that you want to interface with the robot.

What kind of system and application reliability do you expect? Reliability is not the elimination of failures, it is the ability of the system to sense a problem and take corrective action. Not only must the hardware be reliable; overall chemistry must be reliable as well. If the chemistry is unreliable, it will affect the overall application. The weakest link in a system will determine down time.

In summary, expect the following: (1) the system will take several weeks to be operational; (2) you must attend the robotic vendor's school; (3) the application must be planned and time spent on the project; (4) the overall reliability of the system will only be as good as the weakest link; and (5) the robotic system will perform as well as or better than the manual method.

LABORATORY LAYOUT

Automating an application requires a systematic approach to the steps involved in producing a sample for analytical measurement. Before a procedure can be automated, a considerable amount of investigation is required. The procedure should be validated and carefully documented. Human- and chemically-induced variations should be fully understood in terms of how they relate to analytical precision and there should be an additional focus on how the method is performed by an individual—in many cases individual methods differ from the written method.

Planning the robotic system that will best serve the intended application or applications again requires a systematic approach. There may be many decisions that need to be made, even adjustments to a procedure may be necessary before a formal system can be assembled.

In general, automating a procedure involves: (1) setting a goal (determine the number of samples to be processed per x hours of operation time; (2) determining the modules to be used; (3) estimating sample throughput; (4) laying out the benchtop.

Set a Goal

Prior to automating your procedure, expectations for the system should be clearly established. Three such expectations that greatly affect both the procedure and the system are:

1. How many samples are to be processed per day?
2. How long is the robotic system expected to operate per day?
3. Where is human intervention required?

The answers to these questions can have significant impact on the system from the standpoint of the modules and stations used, the program developed to perform the procedure, and even the procedure itself.

If the number of samples and hours of operating time per day must be estimated, a good place to begin is with current manual procedure. Several possibilities may arise if the procedure is being performed by a number of people. You should consider:

1. The number of samples currently being done per day [(samples per person) × (number of technicians doing samples)]
2. Time it takes one technician to do his samples
3. Total time it takes to do all samples done in one day [(time it takes one technician) × (number of technicians doing samples)]

There are several factors that affect where human intervention is required during a sample preparation procedure, and they are often interrelated. To quickly cost-justify a system, human interaction with an automated system should be minimized and the system should run more than eight hours per day. Human intervention affects how long the system can be run unattended. In a single-shift laboratory, a system requiring intervention more than one or two times a day cannot be reasonably run attended on non-shift hours. Different approaches to automating the procedure may have to be considered. These approaches may require a change in instrumentation, laboratory bench layout, or procedure.

The following illustrates the relationship of factor and effect.

Factor	Effect
Number of samples	Size of the lab stations
	Supply of reagents
	Number of disposables
	Size of number of waste receptacles
Size of lab stations	Amount of available bench-top space
	Accessibility of stations

GETTING STARTED IN LABORATORY ROBOTICS

	Number of different stations that can be used in a procedure
End instrumentation	The robot may be able to directly introduce the prepared sample or human intervention may be required
Initial setup	Regular replenishing of supplies by technicians
External instrumentation	Samples must be transported to other instruments in locations other than the current work area for certain preparation steps and a particular number of available samples may be required available to make the job worthwhile

One option that should be investigated is a change from a batch mode of operation, which is typical of the manual procedure, to a serial mode of operation.[16]

The following lists contrast the difference between the manual and serial modes of accomplishing various operations.

Manual procedures

- Samples are often processed in batches because people perform best when doing one task at a time.
- Equipment has been manufactured to be "human friendly."
- Lab stations typically have capacity for a large number of samples allowing the operator to do each step with a number of samples.
- Lab stations have low utilization, i.e., a centrifuge may only be used twice a day.

Serial procedures
- Computers keep track of time and simultaneously control many tasks.
- Samples are processed one at a time and each will have the same history of preparation.
- Lab stations have a much lower capacity in terms of number of samples and take up less space.

Example 1. Sample preparation for HPLC using a batch autoinjector compared to serial/direct injection into the HPLC.[16]

Procedure

1. Four-minute sample preparation. Typical operations include weighing, adding solvent and internal standard, and mixing and filtering, followed by capping a vial and loading samples into an autoinjector (batch) or direct injection (serial).

2. HPLC analysis. Forty-eight samples in batch mode for HPLC autoinjector. Eight minutes per sample.

Timing
Single sample = 4 min + 8 min = 12 min.

Batch
  First sample = (48 min × 4 min) + 8 min = 200 min = 3 hr + 20 min
  Last sample = first sample (batch mode) + (47 min × 8 min) = 576 min = 9 hr + 36 min
Serial
  First sample = single sample = 12 min
  Last sample = first sample (serial) + (47 min × 8 min) = 388 min = 6 hr + 28 min

Summary of example 1

	Batch mode	Serial mode
First sample complete	3 hr + 20 min	12 min
Last sample complete	9 hr + 36 min	6 hr + 28 min
Estimated technician time	3 hr + 12 min plus setup	Setup only

Conclusion. Serializing this procedure provides immediate availability of initial data and completion of the entire sample group more than three hours faster than batch operations. Even if people, working in the batch mode were 50% faster than the robot, serialized results would be completed two hours sooner.

Since the chromatographic time is twice the sample preparation time, one robotic system could feed two HPLC's and double the sample throughput.

Example 2. The following example illustrates batch versus serial approaches for a more complex procedure:

Procedure
1. 3 min pre-preparation (i.e., weigh, add reagent, mix, and load incubator)
2. 60 min incubation
3. 1.25 min additional preparation, manipulation, and loading centrifuge
4. 20 min centrifugation
5. 2.0 min unloading centrifuge and introducing aliquot and sample
6. 10 min analysis

Approach. The total robotic/sample preparation is 6.25 min per sample, and the total analysis time is 10 min per sample. With incubator capacity for six or more samples and centrifuge capacity for two or more samples, the analysis time becomes the rate-limiting element in the serial mode.

Timing

Single sample = 3.0 min + 60 min + 1.25 min + 20 min + 2.0 min + 10 min = 96.25 min = 1 hr + 36.25 min.

24-Sample Batch (requires 24 sample capacity incubator and centrifuge)
    First sample = 72 min + 60 min + 30 min + 20 min + 48 min + 10 min = 240 min = 4 hr
    Last sample = first sample + (23 min × 10 min) = 240 min + 230 min = 470 min = 7 hr + 50 min

Serial
    First sample = single sample = 96.25 min = 1 hr + 36.25 min
    Last sample = first sample + 23 min × 10 min = 326.25 min = 5 hr + 26.25 min

Summary of example 2

	Batch mode	Serial mode
First sample complete	4 hr	1 hr + 36.25 min
Last sample complete	7 hr + 50 min	5 hr + 26.25 min
Required incubator capacity	24 samples	6 samples
Required centrifuge capacity	24 samples	2 samples
Incubator utilization	60 min/batch	Continuous
Centrifuge utilization	20 min/batch	Continuous

There are three major benefits to an automated serial approach:

1. Efficiency. In serialized procedures, laboratory equipment tends to be utilized continuously with samples loaded and unloaded by the robot throughout the procedure. In batch operation, equipment usually has capacity for many samples but is utilized only occasionally.
2. Uniform sample history. In well-planned serialized procedures, each sample experiences identical processing. This reduces or eliminates error sources (e.g., degradation due to variable waiting times) common to batch operations.
3. Faster availability of results. The first sample is available in the time required to process a single sample. Subsequent samples follow at a rate determined by the rate-limiting step in the process. In a batch process all samples within the batch must be prepared prior to the final analysis, which is usually done serially.

Determine Which Modules are to Be Used

Every automation system will include the robot and the system controller. The controller will be used to operate other laboratory station on the benchtop. These laboratory stations, which include the robot, will be used in varying degrees throughout the procedure. The laboratory stations will perform specific lab-scale operations. Many lab stations depend on the robot to bring them samples and later remove them, other stations function entirely independent of the robot.

Of particular concern are the robot-intensive operations. Robot-intensive operations typically have many moves associated with them and usually cannot be performed independently of the robot. Therefore, overlapping of robot-intensive operations may not be possible. These issues become critical when robot speed is seen as affecting sample throughput.

The following list of laboratory unit operations briefly describe the modules that may be used to perform a specific task. In many cases the module can be used in more than one way and, depending on the module chosen, will determine how robot-intensive the operation is.

1. Weighing—Balance, balance interface
2. Grinding—Grinder, grinder interface
3. Manipulation—Appropriate robot hand(s), capping station
4. Liquid Handling—Master lab station, syringe hand, remote nozzles
5. Conditioning—Power and event controller with shaker, vortexer, heating block
6. Measurement—Power and event controller, instrument interface, computer interface
7. Separation—Centrifuge, master lab station for extractions, filtration set up
8. Control—Controller, computer interface
9. Data Reduction—Computer interface
10. Documentation—Printer, computer interface

The robotic-intensive operations are:

- Hand changing
- Sample transport (in vessel or in liquid form)
- Disposables (attaching and detaching)
- Liquid handling
- Capping and uncapping
- Injection

The non-robotic-intensive operations are:

- Conditioning steps
- Weighing
- Data reduction
- Documentation
- Measurement

Estimating Sample Throughput

The process for estimating sample throughput uses six major steps: 1. Outlining the method in detail; 2. determining the robotic manipulation time for one sample; 3. determining the non-robotic times for each lab station and analysis time for one sample; 4. calculating the sample output rate; 5. computing the required sample capacity of the lab stations; 6. computing the input/output sample station capacity.

Outlining the method in detail. A sample preparation worksheet should be prepared and expanded to include a description of each step in the method. Major robotic manipulation steps such as hand changes, pipet pickups, and dispensing operations should be added. In addition, actual parameters such as volumes weights, apparatus, and times should be included. The final step of the procedure should be the analytical technique and its parameters and time.

Determine robotic manipulation time. Robotic manipulations are viewed as "transfers." A transfer is the movement of an object from one location to another on the benchtop and typically involves eight moves. Four moves are used to: (1) go over the object, (2) down, (3) grasp it, and (4) pick it up; four additional moves are used to (1) go to a new location, (2) down, (3) release, and (4) back up. There are also manipulations referred to as "transfer equivalents." Transfer equivalents include such operations as a change of hands and pipet tip attachment and removal.

Assume each move made by the robot consumes approximately 3 sec. A transfer which involves eight moves will take 24 sec. For estimating purposes, allowing 30 sec per transfer is convenient. It also allows some time margin for other locations that will appear in an actual program. Using these time guidelines and transfer definitions, the procedure outlined on the worksheet can be viewed as a series of transfers or transfer equivalents. A conservative estimate of the robotic time can be calculated by totaling the number of transfers and dividing this value by 2 to yield the robotic time in minutes.

Determine the non-robotic time. Each laboratory station may have operations to be performed which do not involve the robot, but which take significant time. Common examples are conditioning steps such as heating, shaking, evaporating, and centrifugation. These steps may take many minutes or hours, and be multiples of the robotic manipulation time. These times do not set a limit on the sample output rate unless the resource involved is extremely expensive and cannot be provided simultaneously for multiple samples. The time required for analysis does set a limit on the sample output rate, however, because most instruments can only analyze one sample at a time.

Set the sample output rate. The sample output rate is the number of samples that can be processed per hour. This rate will be determined by either the robotic manipulation time or the analysis time, whichever is greater. The time that is greater is the rate-limiting step.

Compute station capacity. The sample capacity of the laboratory stations used in the nonrobotic operations can be determined from the sample output rate and the nonrobotic time. The station capacity is the nonrobotic time divided by the sample output rate. This value reflects the number of samples which that station must handle simultaneously. There will be stages during the procedure when the station will not be completely filled. This will occur during the initialization and termination of that particular part of the procedure. An additional of one or two positions may be desired to accommodate sample transfers during the most active phases of that station.

Input/output rack capacity. Real productivity happens with unattended and extended operation of an automated system. This means that human intervention should be minimized. The system should be set up to handle as many samples as possible between known times of operator servicing. The racks or sample holders must therefore be large enough to accommodate all samples that will be processed in this time. Capacity is defined as the intended hours of operation times sample output rate.

Laying Out the Bench Top

After completing the previous three steps, the actual bench layout becomes a task of placing the robot, modules, and racks in efficient locations. Figures 5-1 through 5-8 serve as examples of lab bench layouts.

GETTING STARTED IN LABORATORY ROBOTICS 85

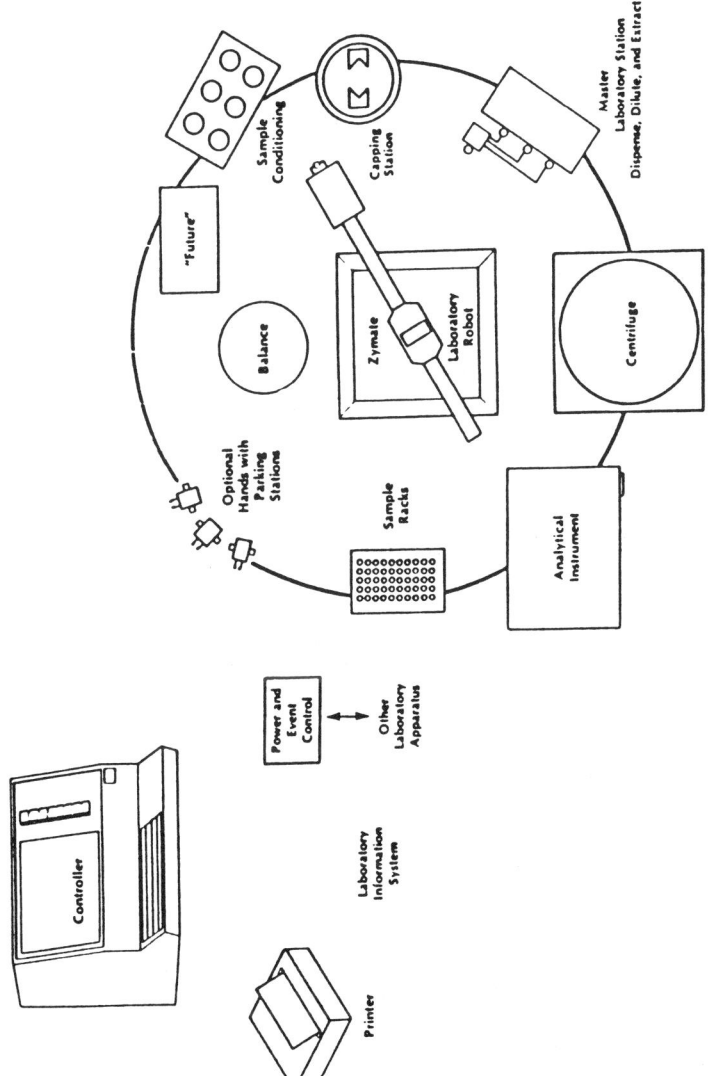

FIGURE 5-1. General robot system layout. (Figure courtesy of Zymark Corporation.)

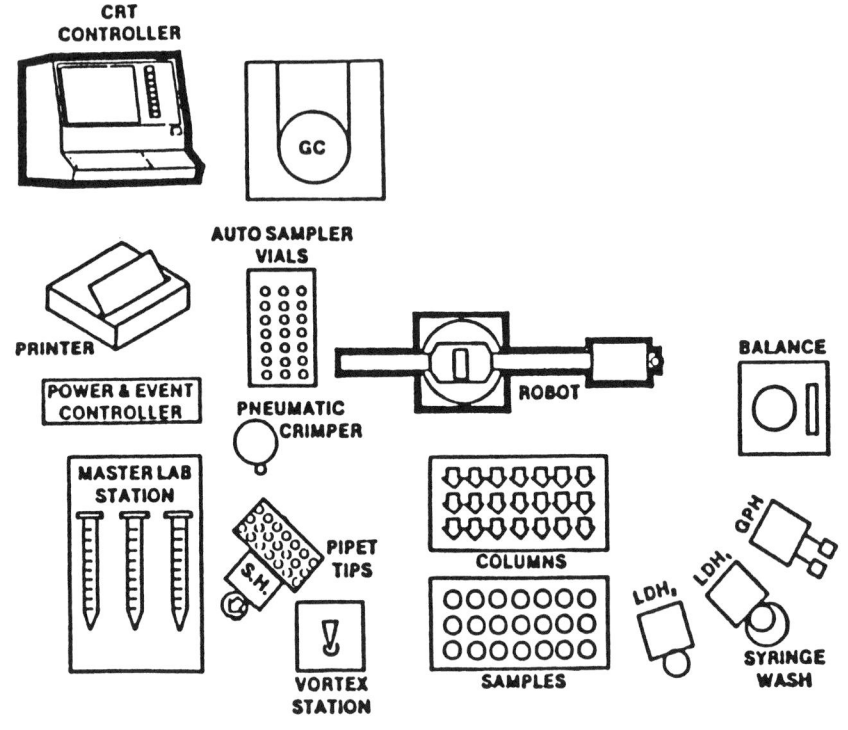

S.H. = SYRINGE HAND
LDH₁ = LIQUID DISTRIBUTION HAND 1
LDH₂ = LIQUID DISTRIBUTION HAND 2
GPH = GENERAL PURPOSE HAND
GC = GAS CHROMATOGRAPH

FIGURE 5-2. Robot system for gas chromatography analysis. S.H., syringe hand; LDH_1, liquid distribution hand 1; LDH_2, liquid distribution hand 2; GPH, general purpose hand; GC, gas chromatograph. (Figure courtesy of Advances in Laboratory Automation—Robotics, G. L. Hawk and J. R. Strimaitis, 1984.)

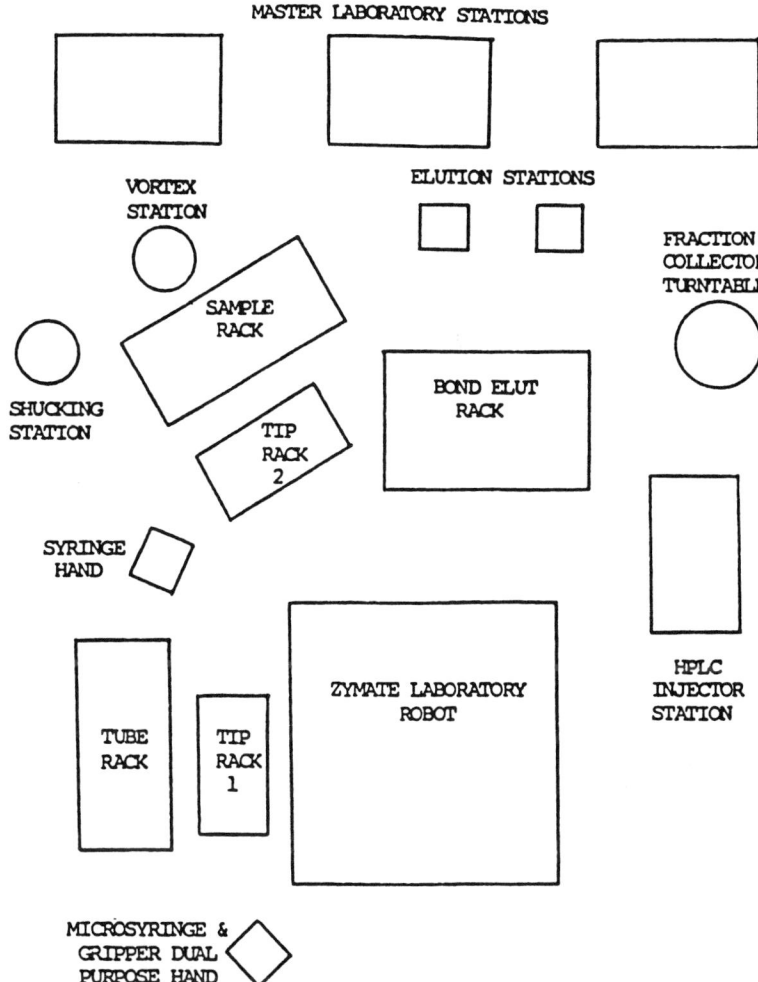

FIGURE 5-3. Robot layout for assays of drugs in animals and human plasma. (Figure courtesy of Advances in Laboratory Automation—Robotics, G. L. Hawk and J. L. Strimaitis, 1984.)

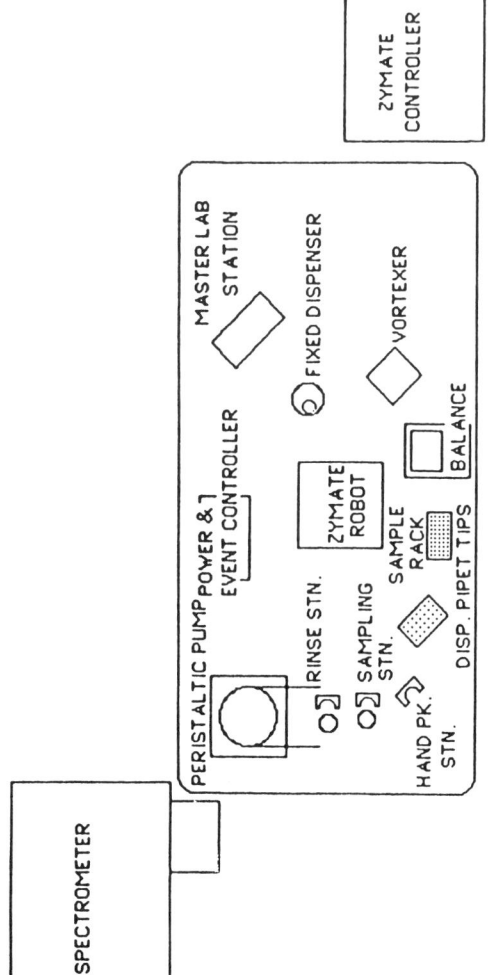

FIGURE 5-4. Robot layout for use with ICP spectrophotometer. (Figure courtesy of Zymark Corporation.)

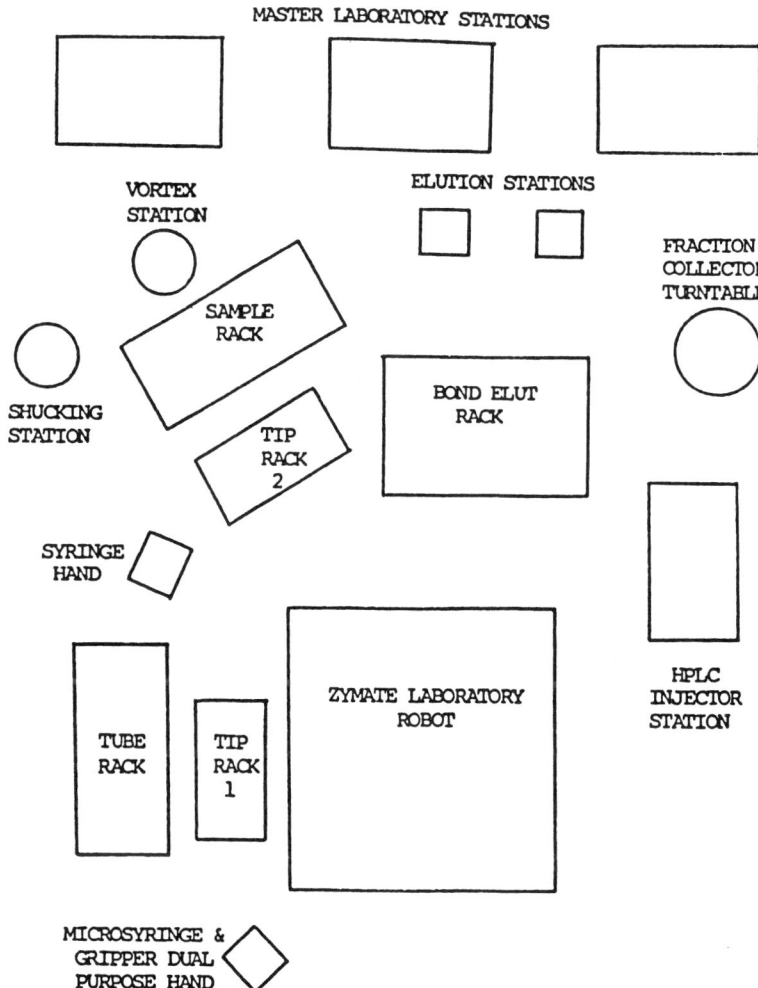

FIGURE 5-5. Robot layout used for HPLC analysis. (Figure courtesy of Zymark Corporation.)

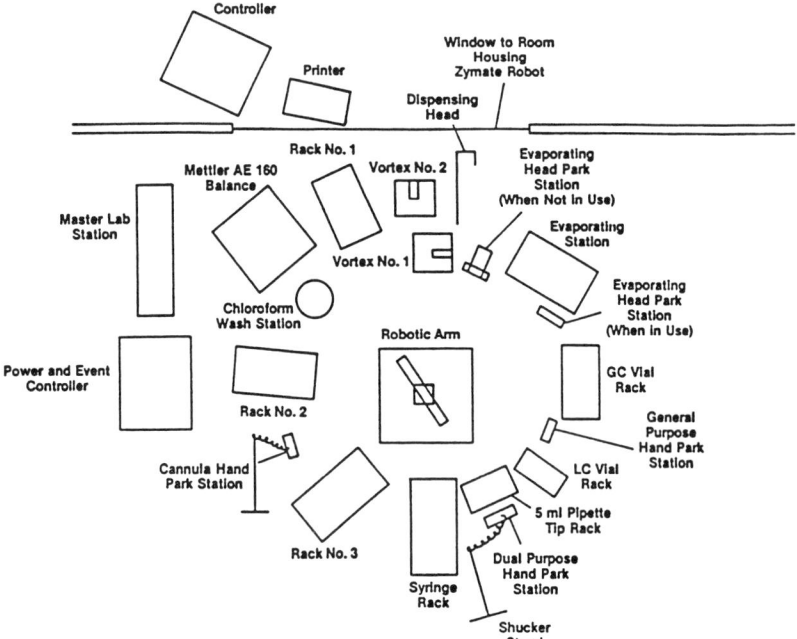

FIGURE 5-6. Robot layout used for pharmaceutical sample preparation. (Figure courtesy of Advances in Laboratory Automation—Robotics, G. L. Hawk and J. L. Strimaitis, 1984)

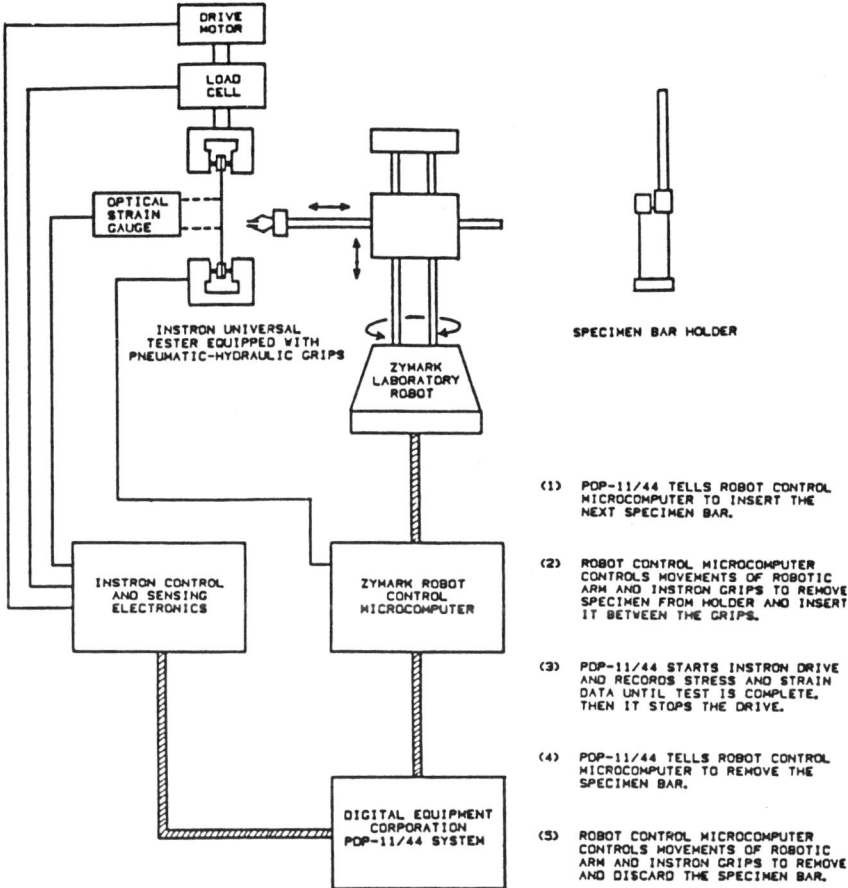

FIGURE 5-7. Robot system automated tensile test. (Figure Courtesy of Advances in Laboratory Automation—Robotics, G. L. Hawk and J. R. Strimaitis, 1984)

92 CHAPTER 5

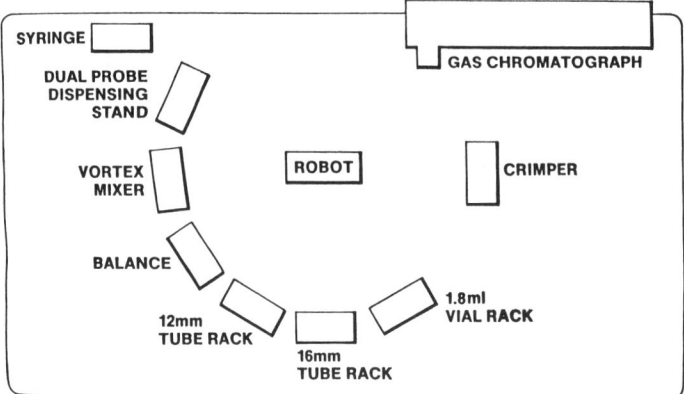

FIGURE 5-8. Robot system layout for gas chromatography. (Courtesy of Perkin-Elmer Corporation.)

6
Programming the Robot

PROGRAMMING THE ROBOT

When a robot finally arrives in a laboratory, it comes with most of its hardware but usually with little software. These terms are carryovers from general-purpose computers and since modern robots are under computer control the use of these terms is reasonable. The concept of a turnkey robotic system for the laboratory is attractive to some scientists but while it would allow the ability to "plug it in" and have it work, it would also destroy the flexibility that is so attractive in a laboratory robotic system.

The programs for the robot tell the robot where to go and what to do at specified times. The term "software" is a catchall term for these programs. Without these programs the robot is limited to a series of commands that govern the general operation of the computer but allow little robotic movement. The mechanism by which these programs are entered into the robot are particular to the robotic system that a scientist chooses to use. For example the Zymate robot uses its own integral computer and keyboard while the Alpha Microbot, Perkin Elmer Master Lab, and Fisher Scientific Maxx 5 use stand-alone microcomputers to perform other functions.

Programming is a series of instructions for the computer or robot and without these instructions the computer or robot is not functional. One soon realizes that without an organized series of instructions the "Garbage In Garbage Out" (GIGO) concept becomes a reality.

A program has four functional activities. The first of these is to get information into and out of the computer. The second is to compare the various operations that the computer has accomplished. In this function, the program will be given a statement that it uses to compare mathematical operations (e.g., if X is greater than Y, then go to Z). This leads to the third function, which is a decision as to what comes next in the program. The fourth function of the program is to instruct the computer to do an operation a defined number of times until a certain condition is satisfied.

The overall function of a program is to convert the solution of the user's problems into instructions for the computer. There are five basic steps to programming:

1. Define the problem
2. Plan the solution
3. Code the problem
4. Test the program
5. Document the program

These steps are critical for all programming, whether it is done for a home computer or a laboratory robot.

In the first of these steps, the user must define the problem. If the user cannot adequately define the problem, then the programming becomes an exercise in futility. The novice programmer is cautioned to define the problem so a reasonable solution can result. If the problem is complex, with a large number of steps, it might be prudent to break it down into a series of simpler operations.

The second of these steps is to plan the solution. It is no doubt evident that if the definition of the problem is important then the planning of the solution is also critical. If the solution is not planned, the programming ends up wandering aimlessly and the desired goal is not accomplished. Professional programmers use a flow chart to accomplish this step of the process. The flow chart is an ordered step-by-step solution. In programming a robot, the user can choose to use a formal flow chart or a more informal approach.

The third step is to code the program. In computer programming, this is a choice of what language to use. There are currently 150 different languages for programming computers. Table 6-1 gives a partial listing of these.

In addition to these various languages there also are a series of prepackaged programs that allow someone interested in running an ap-

TABLE 6-1. Partial Listing of Computer Languages[a]

Language	Use
FORTRAN (*For*mal *Tran*sformation)	Scientific
COBOL (*C*ommon *B*usiness *O*riented *L*anguage)	Business
PL/1 (*P*rogramming *L*anguage *O*ne)	Scientific/Business
RPG (*R*eport *P*rogram *G*enerator)	Business
BASIC (*B*eginners *A*ll Purpose *S*ymbolic *C*ode)	General
FORTH	General
PASCAL	General
LISP	Artificial intelligence applications
ADA	General (especially Department of Defense)

[a]While this table gives some examples of the languages, some languages also have several dialects, so a program written in BASIC for an IBM Personal Computer might not run on a Hewlett-Packard instrument.

plication to accomplish this application without becoming involved in programming the computer. These prepackaged programs may be vendor-supplied, public domain, or purchased programs. A final set of programs are the "user-developed" programs. Programs for robots fall into this category.

There are some general admonitions about robot movements that must be considered when programming a robot. As mature scientists and researchers, we are at times very involved in thinking in abstract and rather involved ways. This tends to be the undoing of novice the novice robot user. If he knew the robot language, a child could easily program robot movement, as he is used to thinking about movements in a more basic manner. When moving an object from one position to another on a bench, an adult realizes that he must avoid objects that would impede his movement (e.g., shelves or other standards), but otherwise he gives little thought to his movements. Unless programmed to avoid objects, a robot will attempt to go through them. This results in wounded pride on the part of the programmer and an appreciation that robot movements must be considered in their most basic form. This can be a most humbling experience.

Communication with the robot via its own internal or external computer isn't direct since the language is a series of 0's and 1's that are not exactly "user friendly," so the series of languages that were mentioned earlier were developed. If a researcher has experience with any of these languages, the ultimate programming of the robot will be simplified.

The Zymark robot, which is the predominant one in use today, employs **Easylab® Software** which is proprietary. This language is user-defined and initially seems similar to BASIC, but BASIC is an unstructured queue of statements. Easylab® is more similar to FORTH, a language in which the elements of a program are stacked upon each other in slices called WORDS. A FORTH program is started by defining a WORD. In FORTH, a WORD refers to an identifiable function or command which in some languages is known as a SUBROUTINE. After execution of the WORD, FORTH waits for the input of a new WORD. The WORD in FORTH may be an arbitrary string of characters. FORTH is a flexible language that consists of a set of standard commands and provides a method to define your own commands which are built on your previous definitions. Easylab® allows the end-user the ability to create his own language and vocabulary which are then stored in the main memory and floppy disc for use or retrieval. In fact, if the robot is a multi-user system, each user could create his own unique vocabulary particular to his own application. Most users of this system have found it reasonable to name positions and programs with a name that typifies that component. For example, a hand with a syringe attached might be called SYRINGE HAND.

The Zymate robot is first taught positions which are stored in the main computer and then these positions are linked into a program to accomplish a specific task. Figure 6-1 gives an example of a Zymate program. Users of this system seem to find it helpful to develop programs about common operations such as weighing or mixing and then combine these in more involved operations. This then allows a better use of computer memory and a programmer's time.

While the Zymark robot is the predominant unit in use in labs today, Owens at Procter and Gamble reported the use of the Alpha Microbot in the analytical lab.[9] This unit was controlled through an external computer. At the 1985 Pittsburgh Conference, Perkin Elmer Corporation introduced a laboratory robot that performs functions similar to the Zymate and Microbot systems. Their system is under the control of an external IBM personal computer and operates under **PERL** (*P*erkin *E*lmer *R*obotic *L*anguage). PERL is the software language developed for use with the Perkin Elmer Masterlab system. The Perkin Elmer system like the Zymate consists of a number of modules such as the robot, and a syringe station or balance. Additionally, there is a system configuration utility which allows you to describe the configuration of the robotic system. Teach programs are a series of programs that allow users to give names to posi-

```
EASYLAB PROGRAM: AGITATE
    ATTHAND1
    HEATER=0
    HAND150
    VOR_CLEAR2
 25 HEATER=HEATER+1
    IF HEATER>3 THEN 50
    BLOCK
    DOWN
    HAND50
    REL0
    VORTEST
    HEATER=HEATER
    BLOCK
    DOWN
    HAND150
    REL0
    GOTO 25
 50 PARKHAND1
```

FIGURE 6-1. Sample Easylab® program.

tions, actions, or quantities. A PERL editor is used to create and edit PERL procedures.[18] Figure 6-2 gives an example of a PERL program.

As was stated earlier, the program is the set of instructions used to operate the instrument. The individual who programs the robot should have some knowledge of BASIC, since the robot languages use similar operations and the programmer will have an appreciation of logical operations necessary to conduct various operations.

Programs can range from simple to exotic and take into account various verification operations to insure that certain tasks are accomplished prior to the execution of other ones. After programs have been written they should be debugged to insure operations flow correctly and smoothly.

The fourth step in programming is to test the program. This testing can encompass several functions. The first and most obvious of these functions is the "desk check" to insure that there are no typographical, mathematical, or logical errors. In robotics the second test is a "dry run" of the program without hands or other peripherals attached. This allows the programmer the opportunity to check the position of the peripherals and to monitor any robot movements without the additional amplification of these associated operations. The final program test done in robotics is running the program using a set of known data. In an automated weighing operation this can be something as simple as using a set of standard

```
procedure demoprog

dim weight(40)
for i = 1 to 25

    test_tubes i

    parallel
        go_to_syringe
        fill_syr_5ml
    end parallel
    dispense_5ml
    go_to_mixer
    mixer_on

    set timer 1 for 15 seconds
    wait for timer 1
    mixer_off
    weight(i) = weigh_sample
    if wt(i) < 0.8 then
        discd_tb
    else
        proc_smp
    end if
next i

end procedure
```

FIGURE 6-2. Example of PERL program for masterlab robot. (Courtesy of Perkin-Elmer Corp.)

weights and comparing the resulting data. A variation of this type of program check would be to run the experiment with the robot using a set of data that was generated from a successful manual method. This test must be done prior to the implementation of the program.

The next part of this chapter will follow the five steps of programming using the Zymart Easylab® language. A similar approach can be followed and implemented regardless of the lab robot manufacturer.

Define the Problem

Develop a program that will allow the automatic weighing of a liquid that will be dispensed by the robot into a test tube. Then add an amount of a second liquid to dissolve the first liquid. The amount of the second liquid will be based on the weight of the first liquid.

Plan the Solution

The robot picks up a test tube and places it on a balance. The test tube is then tared, removed from the balance, and placed into its original position. The robot then moves to a designated location, removes an aliquot of the first liquid, and places it in the test tube. It then removes the disposable tip. Again the robot switches hands, picks up the test tube, and places it on the balance. The weight of the test tube is subtracted from the weight of the tared tube and the sample weight reported. The test tube is then removed from the balance and moved to a nozzle that dispenses the second solution. The amount of the second solution is calculated based on the weight of the sample. The tube containing the solution is mixed for a preset amount of time and replaced in its holder. The unit picks up a third hand which contains the disposable pipet tip with a filter unit attached. This hand can remove and filter an aliquot of the previously prepared solution. The filter can be removed and the solution injected onto an HPLC or placed into a spectrophotometer. After this cycle is completed, it wil be repeated a number of times until all samples are extracted and analyzed or another set of conditions is satisfied.

Code the Program

Figure 6-3 gives a possible sequence of commands that might be used in this sample application, remembering that the language is user-defined. This example shows only one cycle of the proposed application for the sake of simplicity and is only one possible sequence of program events. Each of these commands will then have a series of additional commands nested within it. This example shows the various levels that are possible. Even though this sample application shows only one cycle, additional determinations can be easily accomplished through the use of various conditional statements (i.e., GOTO). The WASH. SYRINGE program is seen in Figures 6-4 and 6-5.

The fourth and fifth steps of program testing an implementation cannot be illustrated but are obvious. Figure 6-6 shows an example of a PERL GC-PREP program.

This type of program checkout allows the programmer to check positions and placement of peripherals prior to actual use. The third and final form of checkout is an actual use in an assay with a set of known data. This includes procedures such as checking of automatic weighing operations with standard weights or a comparing data generated by the robot

SAMPLE/EXTRACTION

TARE.THE.TEST.TUBE

ADD.SOLUTION1

WEIGH.THE.TEST.TUBE

ADD.SOLUTION2

VORTEX

PLACE.TEST.TUBE.IN.RACK

WITHDRAW.SAMPLE

INJECT.ON.HPLC

WASH.SYRINGE

FIGURE 6-3. Sequence of commands for sample extraction program.

WASH.SYRINGE

TO.WASH.STATION

DO.3.TIMES

NO.MODULE.WAIT

DISPENSE.WASH

MODULE.WAIT

FILL.WASH

ENDDO

FIGURE 6-4. Sequence of commands for WASH.SYRINGE program.

COMMAND	EXPLANATION
FILL.WASH	The wash solution that was dispensed is refilled.
ENDDO	A command to indicate the end of a loop started by DO TIMES

COMMAND	EXPLANATION
TO.WASH.STATION	Robot moves to a position above the wash station
INTO.WASH.STATION	Hand containing a syringe or cannula is moved into the wash station
DO.3.TIMES	Do the operation that follows 3 times
NO. MODULE.WAIT	The other peripherals can do tasks without waiting for this one to be completed. This takes these actions out of sequence.
DISPENSE.WASH	Wash solution is dispensed
MODULE.WAIT	The robot is placed back into sequence

FIGURE 6-5a. Explanation of commands for WASH.SYRINGE program.

```
procedure gc_prep
  fill_syr
  disp_liq
  mix_samp
  aliquot
  park_cap
  disp_liq
  crmp_vil
  autosam
end procedure
```

FIGURE 6-6. PERL GC-PREP program. (Courtesy of Perkin-Elmer Corp.)

to data generated by a previous manual method. This is the ultimate test prior to final implementation.

Even if all of these actions have been accomplished the program might harbor some slight inconsistencies, but these will be minor lapses of logic. After the program has been allowed to run for a period of time, the programmer might then make changes to allow the robot to operate more efficiently.

After purchasing a lab robot, those involved must be prepared to invest substantial amounts of time into programming the robot. This time must be in blocks of days and weeks, or implementation will be extremely slow. If this is not done, it will be similar to trying to learn how to use your home computer in 15 minute segments, twice a week. This is not only unproductive but ridiculous.

In late 1986, Zymark introduced a new system architecture dubbed Pyetechnology. This configuration allowed the enduser the ability to set up or reconfigure the system much more easily than was previously possible. Prior to the introduction of this concept, one of the common complaints about laboratory robotics was the extensive amount of time that was required to set up, program, and implement laboratory robotics. The flexibility of laboratory robotics was a double-edged sword that sometimes impeded progress in automation due to the "upfront" time required.

Pyetechnology incorporates the hardware and software necessary to automate a particular laboratory unit operation such as weighting or pipetting. Each of these operations is contained in a Pysection. The Pysection is a wedge-shaped piece that contains the necessary hardware to accomplish the operation of interest. Figures 6-7 through 6-10 give some ex-

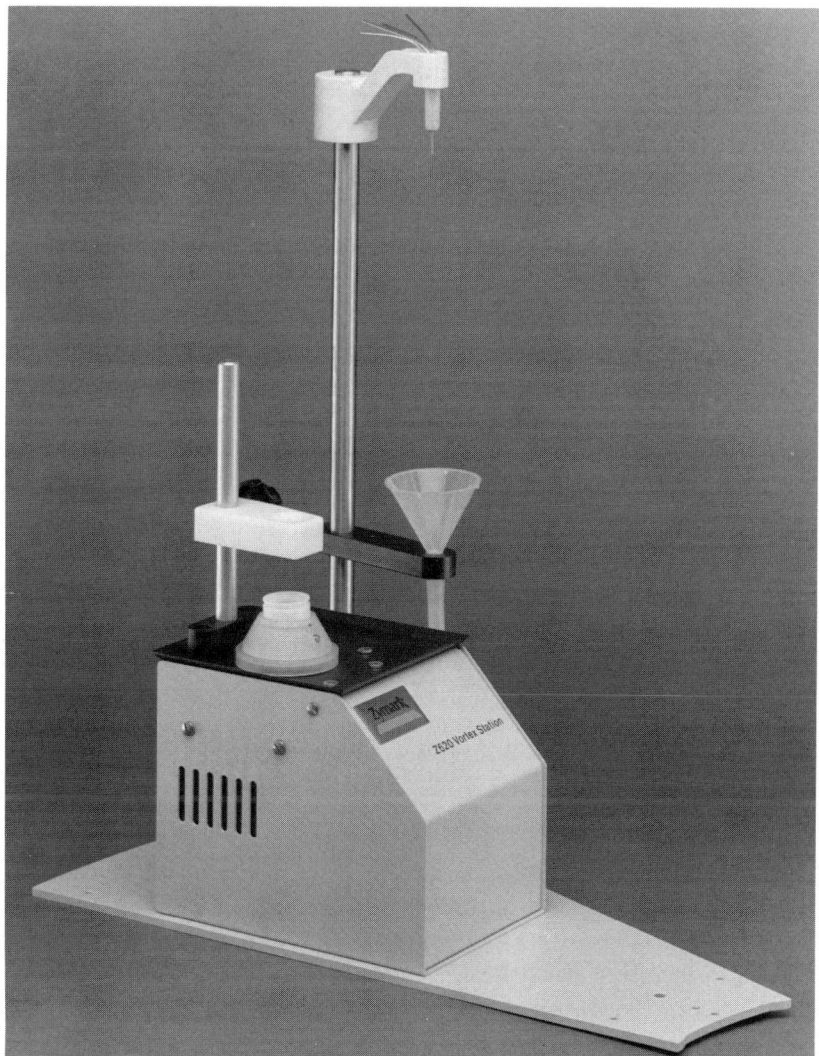

FIGURE 6-7. Dissolve and dilute pysection. (Courtesy of Zymark Corp.)

amples of currently available Pysections. Each Pysection is placed in position around the centrally located robot. These various sections can then be configured to accomplish a specific laboratory task such as sample preparation for HPLC (Figure 6-11).

The user informs the controller of the location of the particular Pysection and its function such as weighing or mixing through the software as-

FIGURE 6-8. Racks for holding sample container. (Courtesy of Zymark Corp.)

sociated with that section. The software contains prenamed commands for that section. Table 6-2 gives an example of a Pysection, and the commands and variables associated with that section. Once the particular sections are loaded and positioned, the user than is required to meld these various sections into the final topdown program.

Currently, the Pytechnology can be applied to the LUO listed in Table 6-3 but one woud expect further sections to become available. This new system architecture will not replace current systems but will add another dimension to lab robotics, allowing accelerated implementations in a series of standard operations.

In summary, the programming of a laboratory robot will involve the use of many skills. An individual must first learn the operation of the robot and its language. The program must be adequately tested and substantial blocks of "quality" time invested. If these objectives are accomplished, then the programming will be just one more successful step in the implementation of laboratory robotics.

PROGRAMMING THE ROBOT

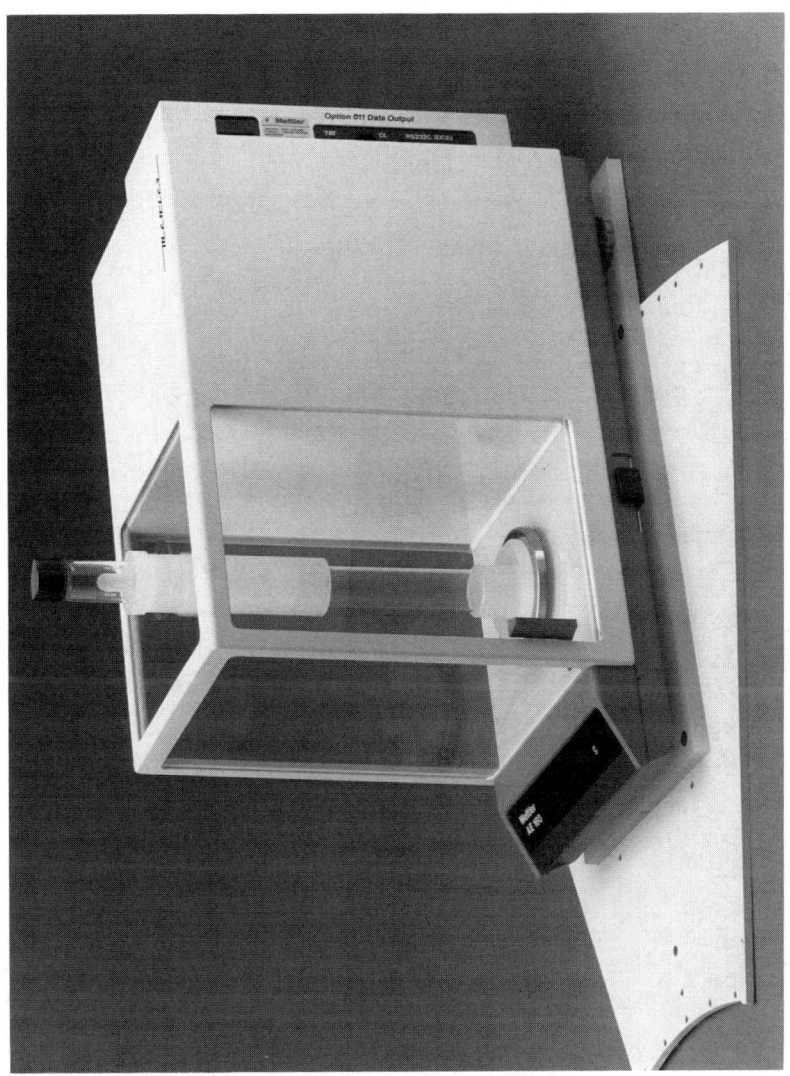

FIGURE 6-9. Weighing station. (Courtesy of Zymark Corp.)

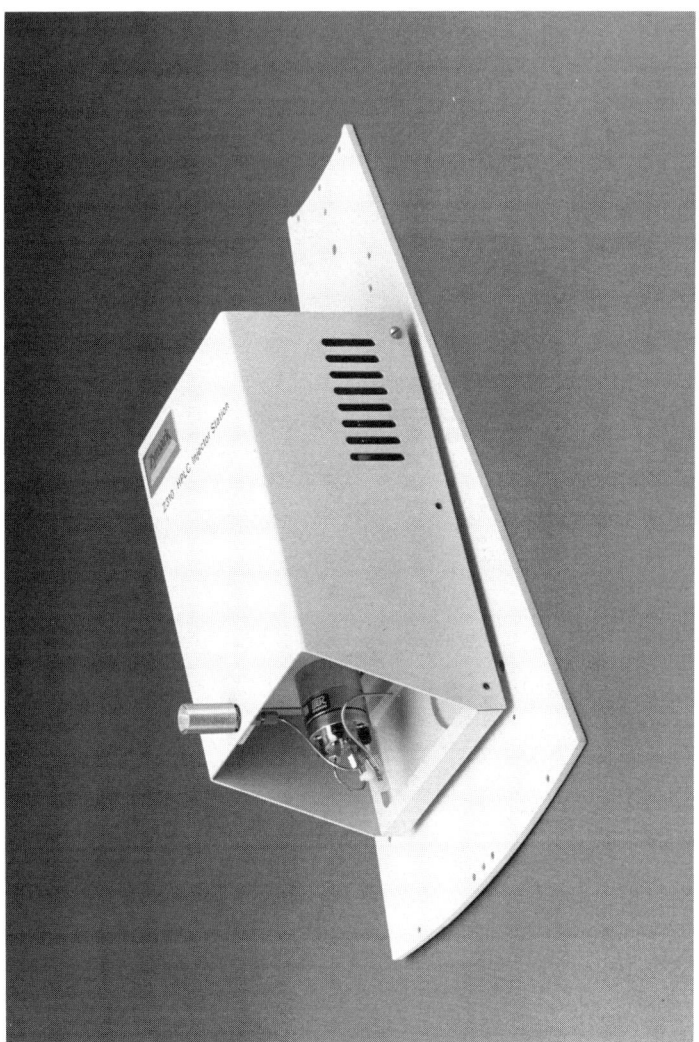

FIGURE 6-10. LC inject station. (Courtesy of Zymark Corp.)

PROGRAMMING THE ROBOT

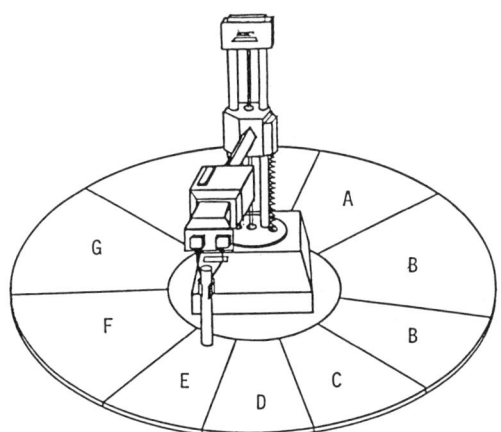

FIGURE 6-11. Sample configuration for HPLC sample preparation. (Courtesy of Zymark Corp.)

TABLE 6-2. Example of Pysection Commands and Variables

Pysection	Command	Variable
Liquid/solid extraction	PUT.CONTAINER.IN.LS.STATION	CONDITION.VOLUME
	CONDITION	CONDITION.TIME
	MOVE.OVER.LS.STATION	ELUENT.VOLUME.1,2, OR 3
	ELUTE.TO.WASTE	ELUENT.TIME.1,2, OR 3
	ELUTE.SAMPLE	
	GET.CONTAINER.FROM.LS.STATION	

TABLE 6-3. Currently Available Pysections

General purpose hand	Linear shaker
Assorted racks	Aspirate and dilute
Pipet/filter with 1 ml pipet tip	Centrifuge
Weighing	Lc inject
Dilute and dissolve	Spectrophotometer-Sip
Crimped vial	Solvent delivery system
Capping	GC inject
Liquid/solid extraction	**Evaporation station**
Vibrating hand	Karl Fisher titrator
Pipet (1-4 ml)	

7
The Future of Laboratory Robotics

What began as a laboratory curiosity about three years ago has now become a laboratory phenomenon. At the Pittsburgh Conference on Analytical Chemistry and Applied Spectroscopy, premier laboratory equipment exposition of 1985, there were more laboratory robots than ever before and more papers on laboratory robotics. In late 1984, the Second International Symposium on Laboratory Robotics drew about 350 scientists from all over the world. The first symposium had about 30 attendees. "Robotics" has become the new laboratory buzzword—so much so that many items that would have been called autosamplers or automated liquid handler in past years are now called robotic sample preparation devices. The questions that now arise concern the future of the laboratory robot and its impact on the modern chemistry laboratory. A discussion on the future must also include comments about laboratory organization, personnel, and considerations relative to the totally automated laboratory.

Current laboratory organizations are organized in any number of manners. In his paper on laboratory organizations, Cook[19] outlined several laboratory organizational structure options. The first of these options is one in which analysis aligned with process and procedures is well documented. This leads to an outline as in Figure 7-1. In a second option, some laboratories are organized according to various products (Figure 7-2), as opposed to being aligned on a process basis which sometimes

Analytical lab

Process A   Process B   Process C

FIGURE 7-1. Laboratory organization along process lines.

results in a duplication of effort. In a third option. laboratories are organized along discipline lines (Figure 7-3) such as organic, inorganic, and physical chemistry. This structure allows more flexibility than the other two but does not focus on the ability to solve problems. Cook also discussed four other options that included laboratory organization for large and small labs.

In the 1980s, there is a possibility of another option that is more highly focused on problem-solving rather than on technique-oriented laboratories.

The impact on laboratory organization will no doubt see the growth of a new breed of scientist in the lab. This person has been referred to as an automation specialist by Zenie[4] of Zymark. Such an individual will be a hybrid of scientist, computer programmer, and lab tinkerer. All of these skills will be necessary because none of these individuals alone would be able to become proficient with lab robotics. The pure scientist is often involved in the accomplishment of esoteric experiments and the contemplation of abstract ideas, and therefore might be unable to appreciate the simplicity of robot operations. The pure computer programmer would be extremely helpful in developing the logic and ultimate programs necessary for efficient error-free operations but would not be familiar with laboratory operations. The final piece of this hybrid, the lab tinkerer, is an extremely important individual. While robotics is continually maturing with the introduction of (OEM) peripherals by manufacturers other than robot manufacturers, each lab has its own unique set of circumstances

Analytical Branch

Product 1   Product 2   Product 3

FIGURE 7-2. Laboratory organization along product lines.

Analytical branch

Inorganic   Organic   Physical

FIGURE 7-3. Laboratory organization by chemical discipline.

that must be considered in the implementation of robotics. These might include the physical layout of the lab or a unique piece of equipment or glassware that might be properly interfaced with the robot. The lab tinkerer can assist in the design of unique devices for robotic operations. If a laboratory doesn't have an individual with all three capabilities then it is helpful to develop a task team approach where individuals with these specific talents are all involved in robot operations.

While robotics will have an impact on laboratory organization and personnel, it is thought that it will also have a substantial impact on the physical layout of the lab. Current instrumental labs have the need for large amounts of bench space but little need for classical laboratory benches. These benches, equipped with gas, air, and vacuum outlets, are expensive and immovable, resulting in a fixed orientation in the lab. A trend that is becoming evident is the use of carts to support various items of equipment with a core of benches available for sample preparation. This results in increased flexibility in the lab where space is a scarce commodity. This trend is continuing in lab robotics with the manufacturer[10] recommending an island in the lab as an ultimate arrangement for the robot with a peninsula arrangement being the second choice. These configurations allow adequate service and operation of the robot with a minimum of inconvenience.

If this trend toward robotics and automation continues, there will be another market area developed for ancillary equipment and expendables for robots. This area started to develop in 1985, when a fume hood designed for use with the Zymark robot was introduced by St. Charles Manufacturing Company (Figure 7-4). No doubt this trend will continue with more manufacturers becoming involved in robotics and automation. Comparison of a certain variety of bonded-phase extraction for liquid/solid extractions with another variety showed that the chosen variety was more "friendly" to the robot. An example of this bonded-phase extraction is seen in Figure 7-5. This disparity is no doubt a temporary phenomenon but none-the-less an important one. It is also reasonable to think that a

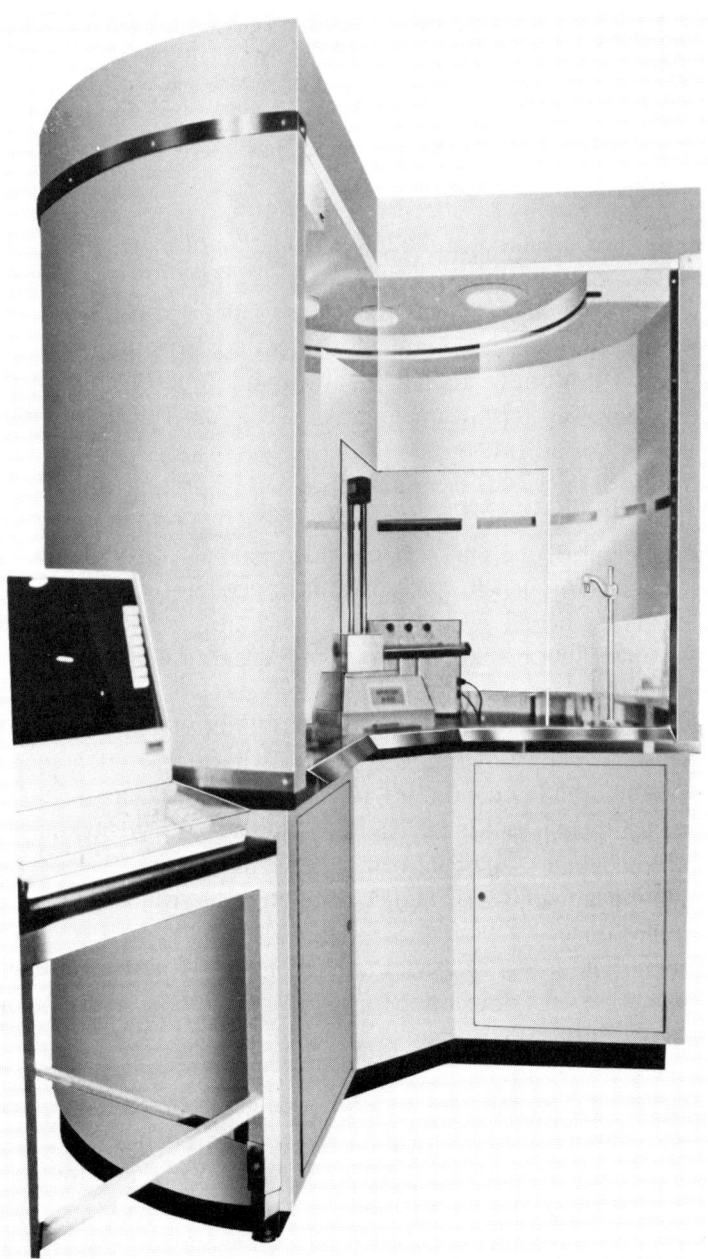

FIGURE 7-4. Hood designed for Zymark robot. (Photo courtesty of St. Charles Manufacturing Company.)

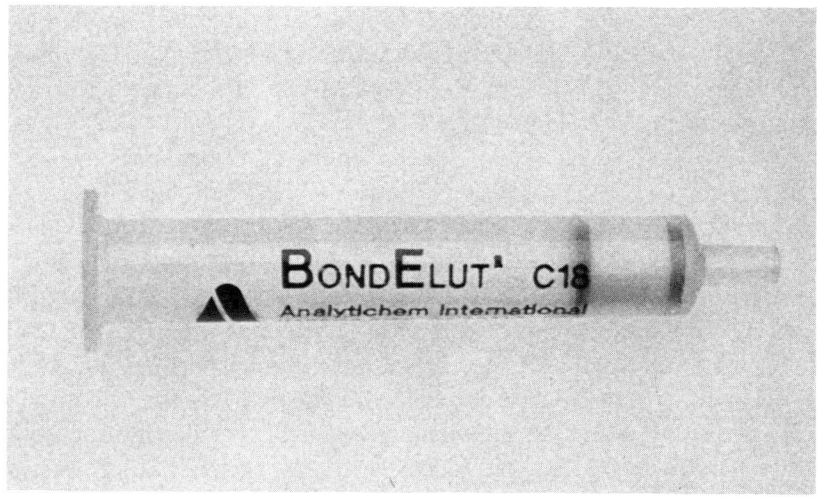

FIGURE 7-5. Bondelut column for liquid/solid extraction. (Photo courtesy of Analytichem International.)

new generation of expendables might be developed for use in automated robotic systems. These items might replace the Erlenmeyer flask and test tube that are commonly found in labs today. This is not to be unexpected, since a new group of expendable items has already been developed for use in many other automated chemistry operations.

Even though the cost of computers has dropped drastically in recent years there is a need to be concerned about their ultimate applicability in a laboratory situation. Two items of concern regarding computer application are evident. The first is the potential noncompatibility of microcomputers associated with various instruments. While many of these are programmable, they are not easily interfaced with other micros or mainframes for purposes of data manipulation or data sharing. The second item is a concern regarding the cost of computers. While the prices of micros have been dropping steadily over the past few years, few corporations can support a wholesale proliferation of different microcomputers which represents a substantial expenditure of capital that might be better dedicated to other resources. In fact, many corporations now have a needs-assessment program that serves to monitor the purchases of microcomputer units.

As robots become more commonplace in laboratories, similar concerns might arise and one might envision a number of laboratory robots that are not totally used. This would defeat one of the purposes of lab robotics—increased productivity.

With these concerns in mind, the possibility of forming a robotic **local area network (LAN)** is extremely feasible. LAN allows independent devices to communicate with each other. The number of devices is limited and the network is under the control of a single organization.[19] A prime example of this type of systems is found in an electronic office environment. At the 1985 Pittsburgh Conference there was an example of a LAN in which an HPLC, FT-IR, and robot were interconnected through the use of a DEC Micro VAX. Additionally, a simple LAN can be currently developed using robots, since a single Zymark controller has the capability to control more than one robotic arm or peripheral. Since some of the lab robots are IBM-PC-based units, a LAN with multiple robots tied together would seem very plausible.

In a similar vein, there is much activity in the interfacing of robots with micro- and minicomputers. Researchers are currently using various microcomputer units to serve as data way-stations to further manipulate and refine data generated through the use of robots. Figure 7-6 shows an example of one of these units.

As the various chromatographic techniques have matured, chromatographic discussion groups have evolved in several areas of the country. These groups encourage an informal exchange of technical information, allow those engaged in this technology the opportunity to meet with others in their field, and develop networks of individuals interested in chromatography. The number of robotic discussion groups has been growing steadily with the growth of analytical robotics. Such groups serve the same function as the chromatographic discussion groups.

Another potential area for the development and refinement of robotics might be in the area of artificial intelligence (AI) or **expert systems**. Expert systems have come into their own in many areas of science. These systems are widely-used computer-based systems. One of the first was developed in 1965 at Stanford and named DENDRAL. Its task was centered on the analysis of organic compounds by mass spectrometry. There are many other expert systems in use in other industries or professions. These include PROSPECTOR for the analysis of geographical data and INTERNIST for the diagnosis of certain diseases. This trend has even made the transition to personal computers as evidenced by a recent advertise-

THE FUTURE OF LABORATORY ROBOTICS 115

FIGURE 7-6. IBM personal computer. (Photo courtesy of IBM Corporation.)

ment in the *New England Journal of Medicine* for an expert system for physicians based on the IBM/PC/XT. It is called PDB (Physician Data Base) Manager.

One can easily envision expert system-robot interactions. In fact the use of CONFIRM or various decision-making steps tends to be the rudimentary start of an expert system. In these techniques, a **microswitch** closure or interruption of a light beam is used to monitor the progress of an experiment. In a simplistic case, a microswitch could be installed on a balance door. If the switch state is monitored then their robot would know whether the balance door was opened or closed. If it detected that the door was closed when it was supposed to be opened it then would not proceed with the remainder of the operation. These expert systems make decisions or help people make decisions. A number of expert systems have been developed that appear to solve complex problems in medicine, science, and engineering. It is plausible to envision the use of an expert system to undertake some decision making in sophisticated robotic systems.

Laboratory robots currently have a set position configured for all peripherals associated with a unit. Through the use of a set of positioning blocks or screws on the table that holds the robot, it is possible to have some latitude in robot configuration but by-and-large these units are dedicated to a single instrument group or assay. If this restriction is ever eliminated, the robot can be totally reconfigured for another operation. Another possible configuration would be the placement of a robot on a series of overhead tracks that would allow its movement to another assay. This concept would impart additional flexibility to a robot-based system, thereby providing another category of experiments where the acquisition of a robot would be appropriate. Figure 7-7 shows a possible configuration using a system of this type.

It is extremely feasible that in the future laboratory robots will be equipped with vision and a tactile sense. These senses might then replace or complement current CONFIRM™ steps. These traits could then be intermingled to form an expert system which would then allow a wide range of robotic operations and cut down on setup and programming time now necessary to achieve robot operations in a laboratory.

Another trend that is evident is the number of cooperative ventures between the robot vendor and other equipment manufacturers. This year a number of these were seen, with the Zymark-HP-Genentech venture as a prime example. One can expect that this trend will continue with each partner bringing their expertise to bear on a problem.

The Fourth International Symposium on Laboratory Robotics was held in October of 1986 in Boston, Massachusetts with over 400 scientists in attendance. There were three days of papers and poster sessions ranging from technical papers on interfacing a robot to particular types of equipment to one of the first papers dealing with the "human issues" of robotics in hospital laboratories. Some of the major events that occurred at the symposium are outlined below. These will not be all encompassing but hopefully will give a bit more information as to the current status of laboratory robotics.

THE FUTURE OF LABORATORY ROBOTICS 117

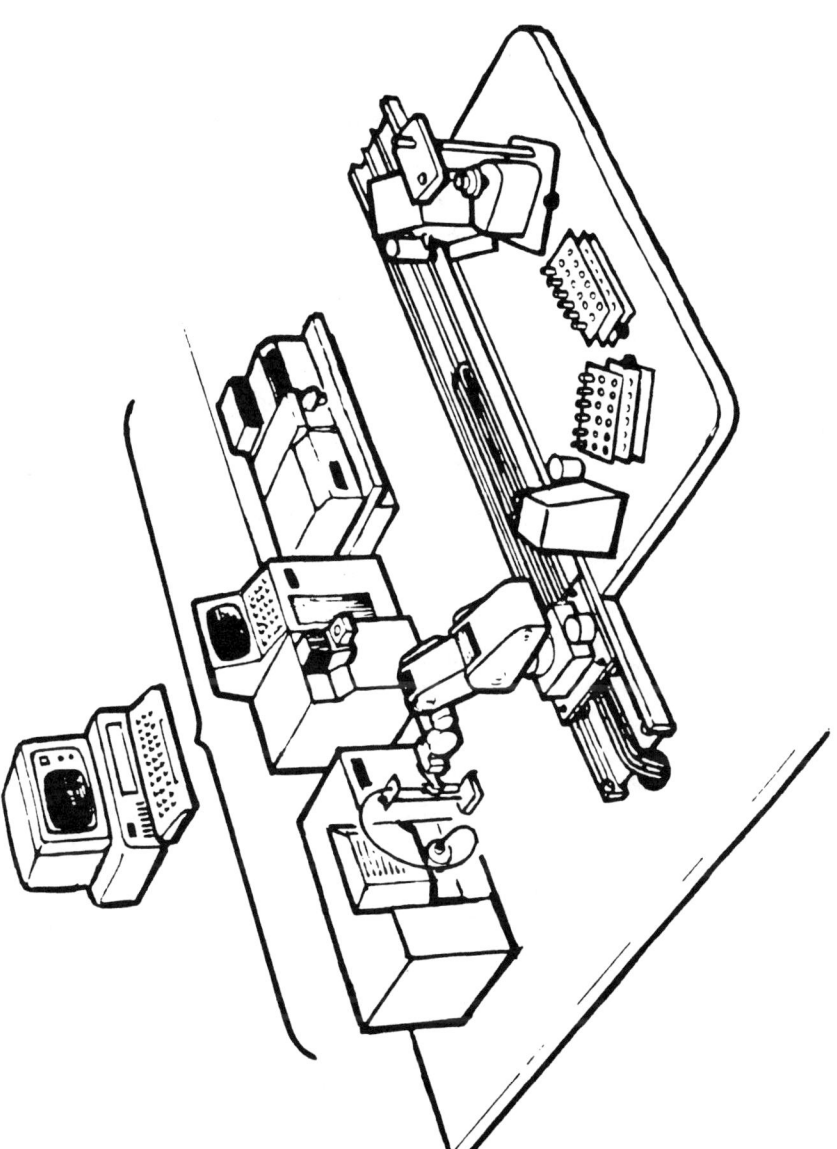

FIGURE 7-7. Robot installed on track allowing for movement. (Courtesy of Perkin-Elmer Corporation.)

Several investigators presented information on the interfacing of robotics with computers and with a plenary session on intelligent robotics. Earlier this chapter discussed some of the possibilities on an expert robot system. Work is well underway with many groups currently active in this area. Continued progress is being made in the area of sample preparation with many organizations acquiring additional robots. Work was also presented in the area of laboratory design with Eli Lilly Corporation presenting a paper on a "robot lab" with flexibility for laboratory operations. One would envision future developments in this area. A study was also presented on a critical comparison of the Zymark and Perkin-Elmer Robotics Systems. The laboratory operations that have been automated continue to cover a wide variety of topics in almost every industry.

This chapter does not begin to touch the surface of the future of laboratory robotics—because an entire volume could be dedicated to that topic. This chapter will hopefully provide some food for thought about the future which is bright and exciting and limitless. It seems that the only limit is your imagination.

Glossary
The Language of Laboratory Robots

The purpose of this glossary is to give definitions of common words and phrases that one will use and come across when dealing in the area of laboratory robotics.[7]

Absolute and relative locations • The ability to define locations to the robot as absolute locations in the work area or as positions "relative" to a previous location.

Algorithm • The step-by-step plan followed by a robot's computer program to make the robot perform some function.

Analog • A continuous electronic signal that reproduces some physical action outside a computer.

Archive • The storage of information for future use.

Articulation • Joints of the robot arm including rotary motion, vertical motion, wrist bend, wrist yaw, radial motion, and wrist swivel.

Artificial intelligence • The science of making a machine do things that would require intelligence if done by humans.

ASCII • A standard encoding of alphanumeric characters into 7 or 8 binary bits.

Asimov's Three Laws of Robotics • The "Three Laws of Robotics" created by Isaac Asimov: (1) a robot may not injure a human being, or through inaction allow a human being to come to harm; (2) a robot must obey the orders given it by humans except where such orders conflict with the first law; (3) a robot must protect its own existence as long as such protection does not conflict with the first two laws.

Automatic buret selector • Allows burets to be selected and moved automatically with the robotic system.

Balance interface • An electronic interface board and software to utilize an electronic balance as a station.

Balance door opener • A device used to automatically open and close a balance door, usually an air cylinder is used.

Bar-coded labels • A readable code on a label that is sensed optically by using a light pen or a laser scanner interfaced to a decoder which transmits the string of characters that is encoded on the label to some external device for display or storage.

Bimba cylinder • A commercially available air cylinder used to actuate a variety of simple jobs to help the robot, such as opening a balance door.

Bit • A single binary digit (0 or 1).

Blank hand • A robot hand for attaching laboratory devices such as a pH probe.

Bug • A mistake or error in the program controlling the robot or other peripheral devices.

Byte • Eight bits of binary data.

Capping station • A laboratory station that is used with the robot to cap and uncap round containers with a screw cap.

Charged-coupled device (CCD) • A special type of computer memory chip which is well suited for capturing pictures and images. CCDs are being used as electronic "eyes" for robots.

Command • An instruction to the robot's computer given in a language or form the computer can understand.

Computer interface • An interface that allows the robot system controller to send data to a laboratory data system.

Confirm techniques • Techniques used in an automation setup to verify that some event has or has not taken place, and taking a preplanned corrective action if a problem has been sensed. The four major types of sensors used as confirm an action are: electromechanical, optical, mass and analytical measurement.

Corrosion resistant robot • A robotic system that has been modified for use in mild corrosive environments. Modification includes replacement of exposed, corrosion susceptible parts with corrosion resistant materials and gas purging lines for the robot electronics.

Central processing unit (CPU) • A special purpose chip that obeys the instruction fed into the computer.

Crash • "Computerese" for a unexpected malfunction which causes a unit to lock up or not operate.

Crimp capping station • A station used to automatically crimp cap vials, such as a GC vial.

Cybernetics • The science of control, how machines and living creatures control important functions and processes, and how they interact with the outside world.

Database • Information available to the robot.

Degrees of freedom • See Articulation.

Dip switch • A set of small switches usually attached to a circuit board that can be used to select various addresses or hardware options.

Dot matrix • A common printer type associated with computers and a method of forming characters and graphics on a display.

Drive system • The source of the robot's power, such as stepper motors, servo motors, or hydraulic power.

Dual function hand • A hand that combines the functions of a general purpose hand and a syringe hand.

EASYLAB® software • A nonstructured and freeform language in which the user teaches the robot and laboratory stations locations and behavior, assigns descriptive names to them, and saves these names in a dictionary for future use.

Evaporation station • A modified block heater with a mounted evaporation manifold. An air cylinder raises and lowers the evaporation manifold to allow the robot to service test tubes in a minimal amount of time.

Expert system • Another name for artificial intelligence.

Feedback • The information received from the outside world regarding some action taken by the robotic system.

Fork • A mounted piece of metal with a "V" cut into it to aid the robot in removing pipet tips and slip-on-caps.

General purpose gripper hand (GP hand) • Provides basic capability to grasp and move vials, test tubes, beakers, and other laboratory devices.

Hydraulic robot • Robot containing hydraulic drive mechanism oil into a cylinder inside the robot, then compressing it with a plunging piston. The increased pressure causes a part of the robot to move.

IEEE-488 • The official numerical designation of the standard for the GPIB (general purpose instrumentation bus).

Instrument interface • A module that provides programmable data acquisition from laboratory instruments and programmable control of other apparatus in the laboratory.

Local area network (LAN) • A communication system that connects a number of computers and their peripherals together to allow information sharing.

Limited sequence robot • A motor starts this robot's arm moving and when the arm gets to where its going it bangs into a preset metal stop. Also known as "Bang-Bang" robot.

Liquid distribution hand (LDH) • A hand that is used for pipeting, manifolding, and remote distribution of liquids from the master lab station.

Loop • Occurs when a robot's computer obeys the same set of instructions over and over.

Manipulator • A robot's shoulder, arm, and hand.

Master laboratory station • A module that provides liquid handling, extraction, and partition capability. It utilizes three computer-controlled syringes to dispense, dilute, and pipet.

Microswitch • A tiny switch used in CONFIRM techniques to verify that either something has happened or it hasn't.

Multichannel pipetor hand • A hand that has eight channels for automatically pipeting up to 250 μl of liquid using disposable pipet tips, typically used in an ELISA Assay.

Optical verification sensor • A simple LED/photocell combination used as an optical sensor to verify the successful completion of an event.

Palletizing • Loading and unloading.

Pneumatic (air-powered) robot • Robot containing pneumatic drive mechanism.

PERL • *P*erkin *E*lmer *R*obot *L*anguage.

PICK-AND-PLACE ROBOT • A simple robot whose only job is to pick something up and place it in another location.

Port • An information outlet connecting a robot's computer to all other devices.

Potentiometer • A variable resister being able to raise or lower a voltage.

Power and event control station • A module that provides programmable control of other apparatus in the laboratory. There are switch closures, input sensors, electrical power outlets, analog/digital converters, and an external DC power supply.

Rack indexing • The ability to define all positions in any uniform rack by teaching only three locations.

Radial motion • The in-out motion of the robot arm.

Random access memory (RAM) • Memory that can be read or written.

Remote dispensing nozzle • A movable nozzle connected to the master lab station that the robot can pick up and use to dispense liquids at remote locations.

Reprogram • The key to a robot's versatility is its ability to learn a new job by programming its computer for a new location.

Robot • Allows computers to do physical work as well as process data.

Robot teaching module • A hand-held module for positioning the robot at locations within its work area.

Read only memory (ROM) • Memory that can be written once and not changed.

RS 232 • A standard for asynchronous serial communications.

Sample conditioning station • A module that can program linear or orbital shaking action and that is designed to stop in a fixed repeatable location so that it can be used in automated procedures with the robot.

Single-shot autoinjectors • Injectors that are used in both GC and HPLC where the robot prepares the sample and then places that single sample into an injector station. While that sample is being run the robot will prepare the next sample.

Slip-on-caps • Caps that are precisely machined to fit over test tubes and other laboratory glassware as an alternative to screw caps.

Solenoid • An electromagnet that is used to activate another device such as a valve.

Syringe hand • A hand that has a motor driven syringe attached to accurately transfer liquids.

Tensile test coupon fingers • Fingers designed for the general purpose hand for use with tensile testing applications.

Tensile test coupon rack • A rack designed to hold rigid "dog bone"-shaped samples for tensile testing.

Titration cup dispenser • This station holds a stack of cups and dispenses them one at a time for the robot.

Titration work station • A station that holds the container during the titration freeing the robot to prepare the next sample while the titration proceeds independently.

Titrator interface • An interface that integrates the communication between the robotic system and the titration unit.

Vibrating general purpose hand • A modified general purpose hand used for pouring powders and other solids to a target weight. The speed of vibration is controlled through the program with feedback from the balance.

Viscous liquid handling kit • Permits the transfer of viscous samples to the Karl Fischer titration cell with minimum exposure to air.

Vortex station • A laboratory station used for single-tube vortexing.

Wrist bend • The up-and-down articulation of the robot's wrist, also referred to as pitch.

Wrist swivel • The rotary motion at the end of the wrist where a hand is connected.

Wrist yaw • The rotary motion of the entire wrist where it connects to the knuckle of the robot arm.

References

1. Asimov, I.; Frenkel, K. A. "Robots—Machines in Man's Image"; Harmony Books: New York, 1985.
2. Information provided by Robot Institute of America.
3. The one-armed chemist, *Forbes* **1985**, *135*, 116.
4. Zenie, F. H. Trends in laboratory automation, *American Laboratory,* **1985**, 51-57.
5. Literature from Zymark Corporation on Laboratory Unit Operations.
6. Liscouski, J. G. Computers in the laboratory, "ACS Symposium Series 265"; 1984; American Chemical Society, Washington, D.C., p 1.
7. Dessey, R., Robots in the Laboratory, *Anal. Chem.,* **1983,** *55*, 1100A-1114A.
8. Siebert, E., An application of robotics to pharmaceutical tablet sample preparation, In "Advances in Laboratory Automation—Robotics"; Hawk, G. L.; Strimaitis, J., Eds., Zymark: Hopkinton, Massachusetts, 1984; p 37.
9. Owens, G. D., and Eckstein, R. J., *Anal. Chem.* **1982**, 54, 2347-2351.
10. Information Provided by Zymark.
11. Justification Guidelines, Zymark Publications, JG 101.
12. Antloga, M.; Markelov, M.; Pagliaro, L., Novel approaches to solving a variety of problems in the industrial analytical chemistry laboratory, In "Advances in Laboratory Automation—Robotics"; Hawk, G. L.; Strimaitis, J., Eds., Zymark: Hopkinton, Massachusetts, 1984; p 137.
13. Scott, R. L.; Rieke, J.K., Development of a fully automated tensile test system, In "Advances in Laboratory Automation—Robotics"; Hawk, G. L.; Strimaitis, J.R., Eds., Zymark: Hopkinton, Massachusetts, 1984; p 154.
14. Mango, P.A., Robotics in polymer testing and characterization, In "Advances in Laboratory Automation—Robotics"; Hawk, G. L.; Strimaitis, J.R., Eds., Zymark: Hopkinton, Massachusetts, 1984; p 17.
15. Kirsch, R. B.; Lang, V. B., The determination of trace levels of polyacrylamide in

water using robotics In "Advances in Laboratory Automation—Robotics"; Hawk, G. L.; Strimaitis, J., Eds., Zymark: Hopkinton, Massachusetts, 1984; p 193.
16. Comparison of Batch and Serial Laboratory Operations, Zymark Literature.
17. Kropscott, B. E.; Dittenhafer, M. K. In "Advances in Laboratory Automation—Robotics"; Hawk, G. L.; Strimaitis, J., Eds., Zymark: Hopkinton, Massachusetts, 1984; p 108.
18. Perkin-Elmer Robot Language (PERL) Instructions (1986).
19. Cook, C. F. Analytical laboratory design in a changing environment, *Anal. Chem.* **1976**, *48*, 724A.
20. Schoenhard, G.; Schmidt, R.; Kosobund, L.; Smykowski, K. In "Advances in Laboratory Automation—Robotics"; Hawk, G. L.; Strimaitis, J., Eds., Zymark: Hopkinton, Massachusetts, 1984; p 64.
21. St. Clair, D. "Network and Communications in Computers in the Laboratory"; American Chemical Society, Washington, D.C., 1984; p 37–44.

Index

Artificial intelligence, 114, 115, 119
Asimov's Three Laws of Robotics, 1, 2, 119
Automation/automated, 1, 22–23, 25, 26, 57–58, 65, 123
Automation, flexible, 2, 11
Autosamplers, 9, 29, 54–55, 60

BOD, 52–53

Capping station, 18, 120
Computer interface, 76, 82, 114, 120

DSC, 54, 65

ELISA, 51, 65

FIA, 11

Easylab, 121

Food/nutrition, 65
Formaldehyde, 59, 60

GC, 28, 55–56, 58, 65, 86, 92
GPC, 53

Hand
  dual function, 44, 48, 82
  general purpose, 48, 82, 121
  liquid distribution, 45, 48, 82, 122
  syringe, 43, 48, 82
HPLC, 11, 28, 29, 65, 89, 103

ICP, 16, 22, 46, 57–58, 59, 88
Instron, 61
Instrument interface, 39, 61, 65, 77, 122

Justification, 25–36
  guidelines, 29–36

Languages, computer, 94, 95, 96, 97
Languages, robot, 96
Laboratory layout, 77, 85–92, 107, 117
Laboratory Unit Operation (LUO), 15, 16, 22, 23, 104
Laboratory organization, 109–111
Liquid/solid extraction, 22–23, 57
Local area network, 114, 122
Lubricating oils, 57–58

Maxx 5, 2, 6, 93
Master laboratory station, 39, 41, 82, 122
Masterlab, 2, 5, 7, 8, 93
Microbot, 4, 8

Nitrogen, 58
NMR, 28, 35–36, 47, 60–61, 65

Organic, 65

PCB, 55
PERL, 96, 97, 99, 102, 122

pH, 61–62
Pharmaceutical, 27, 65, 67, 90
Polyacrylamide, 63, 64
Power and event control station, 39, 40, 82, 123
Productivity, 12, 26
Pysection, 103, 104, 105, 106, 107, 108
Pytechnology, 102, 104

Robot, types, 3, 4, 5, 6, 8

Quality control, 11, 65

Robot, definition(s), 1, 2
Robot friendly, 11, 76
Robot intensive operations, 82
Non-robotic intensive operations, 82, 83–84
Robotic manipulation time, 83

Safety, 11
Sample capacity, 84
Sample preparation, 10, 53–54, 55–57, 64, 65
Spectroscopy, 49–50, 59, 63–64, 65

Tensile testing, 61, 91
Thin layer chromatography, 56, 57, 65
Titration
  pH ISE, 31–33, 49, 53
  Karl Fischer, 48, 62–63
Total solids, 52
Turnkey system, 73

X-ray fluorescence, 64, 65

Water, 63

Zymate, 2, 93, 3, 8, 31, 96